Particle Technology Series

Volume 24

Many materials exist in the form of a disperse system, for example powders, pastes, slurries, emulsions and aerosols, with size ranging from granular all the way down to the nanoscale. The study of such systems necessarily underlies many technologies/products and it can be regarded as a separate subject concerned with the manufacture, characterization and manipulation of such systems. The series does not aspire to define and confine the subject without duplication, but rather to provide a good home for any book which has a contribution to make to the record of both the theory and applications of the subject. We hope that engineers and scientists who concern themselves with disperse systems will use these books and that those who become expert will contribute further to the series.

The Springer Particle Technology Series is a continuation of the Kluwer Particle Technology Series, and the successor to the Chapman & Hall Powder Technology Series.

More information about this series at http://www.springer.com/series/6433

Colin Thornton

Granular Dynamics, Contact Mechanics and Particle System Simulations

A DEM study

Springer

Colin Thornton
School of Chemical Engineering
University of Birmingham
Birmingham, UK

ISSN 1567-827X
Particle Technology Series
ISBN 978-3-319-18710-5 ISBN 978-3-319-18711-2 (eBook)
DOI 10.1007/978-3-319-18711-2

Library of Congress Control Number: 2015949154

Springer Cham Heidelberg New York Dordrecht London

Printed on acid-free paper

Springer International Publishing AG Switzerland is part of Springer Science+Business Media (www.springer.com)

Foreword

Our crowded bookshelves all contain just a few 'essential' books – the ones we tend to reach for as the most authoritative source on whatever the subject is. This book is destined to become the essential text on the Discrete Element Method (DEM).

It is well known that the majority of products of the chemistry-using industries are either sold as, or pass through, a particulate or dispersed form. Modelling of these dispersed multiphase systems is therefore of economic importance. But that is not the author's main motivation. What he sets out to do is to understand the often unexpected emergent behaviour of these systems – whether useful or not – by building up models from the fundamental physics of their interparticle interactions.

DEM is deceptively easy. All that is apparently necessary is a framework for building up a set of particles (or some other dispersed form such as bubbles or fibres or people), some laws of interaction between them, and a means of stepping through their developing behaviour, keeping track of forces and the resulting movement of the particles at each step. However, each part of the process can be accomplished superficially or thoroughly. Arguably, the real value of DEM lies not in prediction of the outcome of equipment design variations, for example, but in 'numerical thought experiments' where investigators can ask 'what if' questions just as in real experiments. But for such 'experiments' to be useful, the fundamental physics behind the DEM simulation must be correct. Rigour in understanding the contact behaviour between particles has been one of the author's major contributions to the subject, and it is extremely welcome that he has assembled the work of a research lifetime on this subject here in one place. Contact behaviour includes all the complex variety of surface physics, including adhesion, friction, elasticity and plasticity, usually all interacting at once in a complex geometrical arrangement. The author has battled over decades with these problems and includes here every permutation which is found in current DEM practice.

DEM is not just about setting up appropriate contact equations and operating a calculation framework with 'good housekeeping'. It is also about making the most out of the results which are obtained and drawing appropriate conclusions from them. Since the volume of data is so much more than in most real experiments, the post-experimental analysis presents far more choices for the researcher. Here again, the author has spent decades thinking about how best to go about these choices and how to present the results in the most revealing way. This book gives us the benefit of that experience.

It is always helpful for a foreword to give a glimpse of the personality behind the author. Colin Thornton trained as a civil engineer, including site work in all weathers. He therefore came to particle technology through interest in soils, perhaps the most complicated of particulate systems. He realised that the theories for behaviour of such systems were at best semi-empirical and wanted to do better. The advent of DEM, through the work of Cundall and Strack, gave him the opportunity he needed to develop a more fundamental understanding of what was going on, and he has been relentless in pursuing this direction ever since.

Colin is above all a persistent man who (if I can put it this way) doesn't suffer fools or foolish explanations gladly. In other words, he doesn't accept easy answers or 'received wisdom'. Like all the best scientists, he works things through for himself and is reluctant to take shortcuts. I have had the pleasure first of being external examiner to some of his PhD students when he worked at the University of Aston and then of working with him as a colleague at the University of Birmingham. In particular, we worked together on some aspects of fluidisation, which was a new area for Colin. As a reasonably experienced researcher in fluidisation, I was able to explain quite a lot of the accepted theory and practice to him. Nevertheless, the work went slowly because he insisted on questioning every point! He wanted, as ever, to build his own thorough understanding of what was going on, from the ground upwards. I felt challenged, in the best scientific sense. The reader will be able to see some of the results of that work in the present volume. I am sure that everyone Colin has worked with will have considered themselves similarly challenged, and their work has been the better for it.

Guildford, UK
17th May 2015

Jonathan Seville

Preface

This book is about the author's personal obsession with the complex behaviour of particle systems. It is not intended as a textbook or a manual on how to do DEM simulations. It is merely a collection of the author's contributions to particle system simulations. As such it should be of interest to other DEM researchers, but it is also hoped that it will be informative to others working in particle technology/process engineering and geomechanics, both experimentalists and theoreticians, in academia and in industry.

My initial inspiration for studying granular media at the particle scale was during my PhD studies when I was introduced to the papers of Prof. Peter Rowe on 'stress dilatancy theory'. However, it wasn't long before I realised that, in the context of sand, this was no more than an empirical correlation that was applicable under certain limited experimental conditions. This led me to consider regular arrays of spheres and here I was inspired by the papers of Mindlin and coworkers, which included analytical solutions for the contact interactions of elastic spheres, and the paper by Rennie (1959), who considered rigid spheres.

The next logical step was to consider random arrays of polydisperse systems of spheres, and in this area I am indebted to the late Prof. Gerard de Josselin de Jong who explained to me how one could calculate stress and strain for particle systems and the late Prof. Touran Onat who taught me everything I know about microstructural anisotropy, i.e. fabric. However, I am most indebted to Dr. Peter Cundall who, in late 1979, gave me a copy of his 2D DEM code BALL and, in 1987, gave me a copy of his 3D code TRUBAL on which all the results of the simulations presented in this book are based. Without Peter's generosity this book would not exist.

The book, in effect, tells the story of the author's personal experience of DEM simulations. Although trained as a civil engineer, in the mid-1980s the author experienced difficulties in obtaining funding in soil mechanics. One referee to a research proposal even suggested that this (DEM) was not fundamental research! Fortunately the author discovered particle technology where it was deemed that the most important research need was to know the nature and magnitude of the forces at the contacts between particles.

Acknowledgments are due to all the PhD students and post-docs that I have supervised who, of course, were the ones who actually did the hard work. I also thank many other fellow researchers with whom I have had detailed research discussions over many years, especially Prof. Mike Adams. However, most importantly, I need to recognise the contributions of Dr. David Kafui who, as my senior post-doctoral fellow for 18 years, funded by 3-year contracts, was the corner stone of my research group. David provided the initial training and subsequent day to day advice and assistance to all my research workers, and, without him, the all-important progressive continuity of the research group could not have been accomplished.

Finally, I would like to thank my family, especially my wife Margaret, who persuaded me, at the age of 25, to go to university and who financially supported me during my undergraduate and postgraduate studies and, ever since, has always been there for me.

Birmingham, UK Colin Thornton
31st March 2015

Contents

1 Introduction 1
1.1 Origins 2
1.2 Early Work 3
1.3 Outline of the Book 9
References 10

2 Theoretical Background 13
2.1 Granular Dynamics 13
2.1.1 Particle Kinematics 13
2.1.2 Contact Forces 14
2.1.3 Timestep 16
2.1.4 Damping 17
2.2 Assembly Mechanics 18
2.2.1 Stability Considerations 18
2.2.2 Microstructure 19
2.2.3 Stress 21
2.2.4 Assembly Modulus 23
References 25

3 Contact Mechanics 27
3.1 Elastic Interactions 27
3.1.1 Normal Interaction 28
3.1.2 Tangential Interaction 30
3.1.3 Contact Moment 38
3.2 Elastic-Plastic Interactions 39
3.2.1 Normal Interaction 39
3.2.2 Tangential Interaction and Contact Moment 45
3.3 Adhesive, Elastic Interactions 46
3.3.1 Normal Interaction 47
3.3.2 Tangential Interaction 49

3.4 Adhesive, Elastic-Plastic Interactions 51
3.4.1 Normal Interaction 51
3.4.2 Tangential Interaction 54
References 55

4 Other Contact Force Models 57
4.1 Linear Spring Models 57
4.2 Non-linear Spring Models 60
4.3 Partially Latching Spring Models 61
4.4 Adhesive Piecewise Linear Models 63
References 69

5 Particle Impact 71
5.1 Normal Impact 71
5.1.1 Elastic Impact 72
5.1.2 Effect of Plastic Dissipation 73
5.1.3 Effect of Surface Energy 75
5.2 Oblique Impact 77
5.2.1 Rigid Body Dynamics 80
5.2.2 Dimensionless Groups 82
5.2.3 Effect of Elasticity 82
5.2.4 Effect of Plastic Dissipation 84
5.2.5 Effect of Initial Spin 88
5.2.6 Effect of Surface Energy 90
References 90

6 Agglomerate Impacts 93
6.1 Agglomerate-Wall Impacts 94
6.1.1 2D Simulations 94
6.1.2 3D Simulations 99
6.2 Agglomerate-Agglomerate Collisions 118
References 120

7 Fluidised Beds 123
7.1 Theoretical Considerations 124
7.1.1 Fluid-Particle Interaction Force 124
7.1.2 Particle-Fluid Interaction Force 125
7.2 2D Simulations 126
7.2.1 The Transition from Fixed to Bubbling Bed 128
7.2.2 The Transition from Bubbling Bed to Turbulent Bed 130
7.2.3 The Transition from Turbulent to Fast Fluidisation 131
7.2.4 Effect of Surface Energy 133
7.3 3D Simulations 138
7.3.1 Bubble Formation 140
7.3.2 Bubble Rise 142
7.3.3 Bubble Splitting 143
References 145

8 Quasi-static Deformation 147
8.1 Failure Conditions for Regular Arrays of Rigid Spheres 148
8.1.1 Solutions 150
8.2 2D Simulations 154
8.2.1 Direct Shear Tests 155
8.2.2 Shear Bands 160
8.2.3 Simple Shear 164
8.3 3D Simulations 172
8.3.1 Axisymmetric Compression 173
8.3.2 General 3D Stress States 183
References 193

List of Colour Figures in Printed Book

Chapter 1: Fig. 1.9
Chapter 5: Fig. 5.13
Chapter 6: Fig. 6.30
Chapter 7: Figs. 7.1, 7.4, 7.5, 7.9, 7.10, 7.12, 7.13, 7.16, 7.18, 7.19 and 7.20

Chapter 1
Introduction

Abstract Particle systems comprising assemblies of individual grains or agglomerates are widespread in nature and in industry. When examined at the macroscale, they may appear to behave like a solid, a liquid or a gas depending on such factors as the phase volume of the particles and the flow characteristics of the superficial gas. Experimental research has normally been limited to measurements at the macroscale and attempts to provide a rational explanation of the macroscopic phenomena observed have relied on intuition. In order to better understand the behaviour of particle systems numerical modelling techniques have also been used. Traditionally, continuum modelling has treated particle systems as complex pseudo-solids or complex pseudo-liquids with different highly complicated constitutive equations/ closure relationships depending on the state of the system. However, with continually increasing computer power it is possible to treat particle systems as discontinua, recognising that sands, powders and grains are composed of discrete particles that interact with each other at the microscale. This discontinuum approach is now widely used in many areas of science and engineering and is commonly referred to as the Discrete Element Method (DEM). This chapter discusses the origins of particle system simulations and presents some early applications of DEM.

Particle systems comprising assemblies of individual grains or agglomerates are widespread in nature and in industry. When examined at the macroscale, they may appear to behave like a solid, a liquid or a gas depending on such factors as the phase volume of the particles and the flow characteristics of the superficial gas. Experimental research has normally been limited to measurements at the macroscale and attempts to provide a rational explanation of the macroscopic phenomena observed have relied on intuition. In order to better understand the behaviour of particle systems numerical modelling techniques have also been used. Traditionally, continuum modelling has treated particle systems as complex pseudo-solids or complex pseudo-liquids with different highly complicated constitutive equations/ closure relationships depending on the state of the system. However, with continually increasing computer power it is possible to treat particle systems as discontinua, recognising that sands, powders and grains are composed of discrete particles that interact with each other at the microscale. This discontinuum approach is now widely used in many areas of science and engineering and is commonly referred to as the Discrete Element Method (DEM).

C. Thornton, *Granular Dynamics, Contact Mechanics and Particle System Simulations*, Particle Technology Series 24, DOI 10.1007/978-3-319-18711-2_1

The Discrete Element Method (DEM) is a numerical simulation technique appropriate to systems of particles in which the interactions between contiguous particles are modelled as a dynamic process and the time evolution of the system is advanced by applying a simple explicit finite difference scheme to obtain new particle positions and velocities. The technique can be used both for dispersed systems in which the particle-particle interactions are collisional and compact systems of particles with multiple enduring contacts. Consequently, although particle systems may have the superficial appearance of behaving like a gas, a liquid or a solid when observed at the macroscopic scale, all these different states can be investigated using DEM. In principle, DEM can be used to (i) examine data that is normally inaccessible (ii) perform rigorous parametric studies (iii) explore technical innovation prior to prototype testing. DEM results can be used to develop new continuum models that link the information at the micro-, meso- and macro-scales (the academic holy grail); directly relate the macroscopic performance to the intrinsic properties that define the particle specifications; provide detailed information about what happens inside particle systems (which may include new information – discoveries) and thereby provide a rational framework for decision making that can impact upon our everyday thinking.

1.1 Origins

It could be argued that particle system simulations started with the publication of the paper by Metropolis et al. (1953) in which two dimensional systems of hard discs were examined using a Monte Carlo approach to study systems in thermodynamic equilibrium. This led to the development of Molecular Dynamics (MD) in which individual particle trajectories were calculated as they interacted with their neighbours, Alder and Wainwright (1959). In MD calculations the forces between particles are simply a function of the distance between them and the particles are frictionless and consequently do not rotate. For larger particles, i.e. micron-sized and above, more complex interactions are necessary and this led to the application of Granular Dynamics (GD) to rapid granular flows in the early 1980s (Walton 1983, 1984; Campbell and Brennen 1983, 1985). In GD simulations the particles are considered to be rigid and collisions are instantaneous. Energy is dissipated during collisions due to surface friction and coefficients of restitution. Simulations of rigid spheres are ‘event driven’ in that the time step used to advance the simulation is the time to the next collision. This varies during a simulation and at the limit when there are simultaneous or continuous contacts, as in dense compact systems, the time step goes to zero and the GD approach is no longer applicable. Hence, the GD technique is generally restricted to relatively dispersed systems subject to rapid granular flows. For more details of the GD approach the reader is referred to the book by Pőschel and Schwager (2005). An alternative rigid body approach that can be applied to quasi-static deformation of compact systems with enduring contacts was developed by J.J. Moreau and is known as the Contact Dynamics (CD) method; see Chapter 2 in Radjai and Dubois (2011).

An alternative approach to simulating particle systems, sometimes referred to as the 'soft sphere' approach, originated with the work of Cundall (1971). In their seminal paper, Cundall and Strack (1979) describe the technique as the distinct element method but it is now more commonly referred to as the Discrete Element Method (DEM). In the DEM approach collisions occur over a finite time during which the contact forces evolve as the relative approach of the two particle centroids varies during compression and restitution. In the original paper, Cundall and Strack (1979), the particles were allowed to overlap each other at the contacts to represent the local deformation of the individual particles which were considered to be connected by linear springs in the normal and tangential directions. The normal and tangential contact forces were calculated incrementally as the product of the spring stiffnesses and the relative surface displacement increments of the two contacting particles. As a consequence, the computational technique is applicable to quasi-static deformation of compact systems with multiple enduring contacts, rapid granular flows that may be dominated by collisions and heterogeneous flows in which both collisions and enduring contacts coexist.

1.2 Early Work

Cundall (1971) presented two computer codes (i) the BLOCK code for applications to rock mechanics and masonry walls and (ii) the BALL code for applications to soil mechanics and particle technology. Figures 1.1 and 1.2 illustrate the first applications of these two codes. Figure 1.1 shows a configuration of square blocks that represent a blocky rock system. The shaded blocks are fixed with the remainder free to move in any direction. With an interface friction angle set to a low value of 1° the blocks were allowed to settle until an equilibrium state was approached. The interface friction angle was then set to 20° and seven of the fixed (shaded) blocks were removed. As can be seen from the figure, the result was failure by toppling. Figure 1.2 shows a triangular pile of close packed discs that is initially allowed to consolidate under gravity using a high interface friction. Then, the interface friction angle was reduced to 10° and, as shown in the figure, the pile collapsed.

Following the publication of Cundall and Strack (1979) two conference papers were published Cundall et al. (1982), Cundall and Strack (1983). For the benefit of soil mechanics readers it is worth quoting the following observations on the micromechanics of granular material during quasi-static deformation: "Contact forces are concentrated in stiff chains of particles." "Slip almost never occurs in the stiff chains, but in the relatively unloaded regions between chains." "Observed macroscopic velocity discontinuities do not consist of contiguous lines of slipping contacts, but often correspond to 'hinge' regions, involving coherent particle rotations (spins)." "Particle spins contribute largely to deformation in an assembly; the stiff columns collapse in a 'buckling' mode that involves spins of alternating

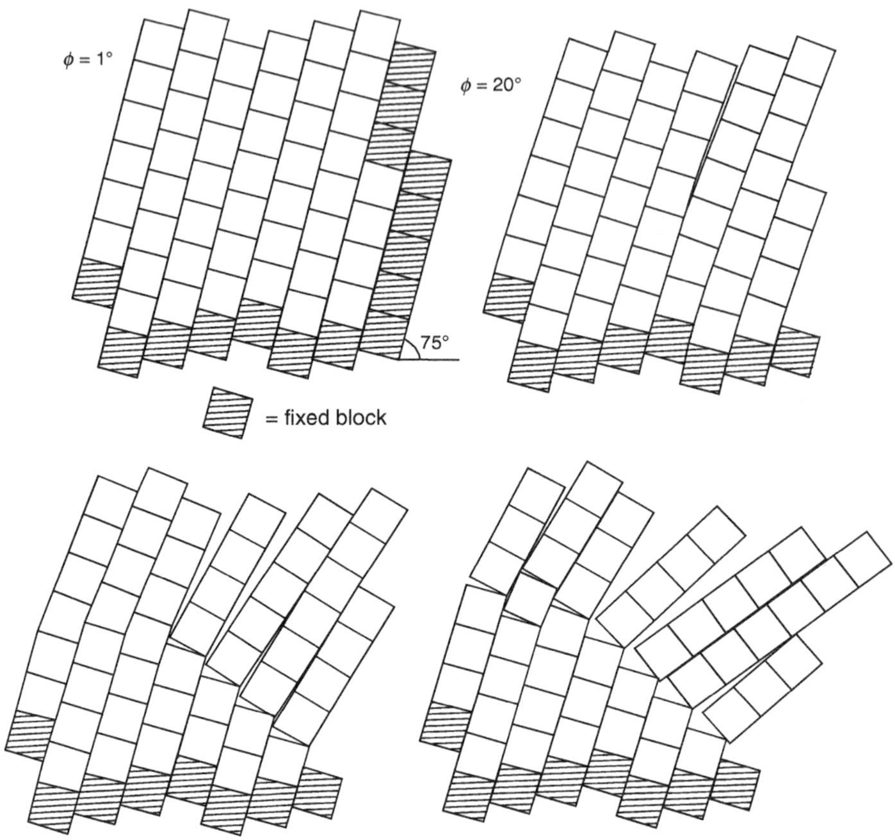

Fig. 1.1 The first application of the BLOCK code (Cundall 1971, Figs. 3–6)

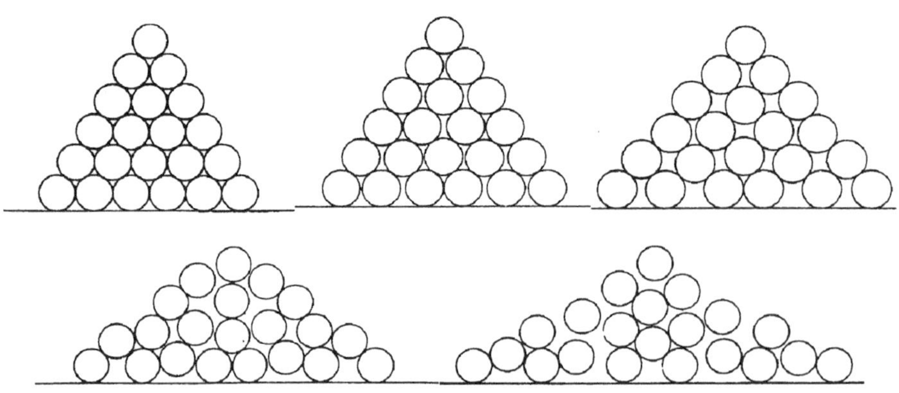

Fig. 1.2 The first application of the BALL code (adapted from Cundall 1971, Fig. 2)

sign." "During deviatoric loading, contacts with normal near the minor principal direction are broken preferentially, and are not fully recovered on unloading." "After a sample undergoes some sequence of loading and unloading, 'locked-in' shear forces persist at contacts even though the boundary deviatoric stress is zero."

In the 1980s DEM simulations were normally run on mainframe computers in a time-sharing environment. This was extremely tedious and time consuming. For example, Thornton and Barnes (1986) reported quasi-static shearing of a polydisperse assembly of 1000 discs. The simulations were performed on a CDC7600 'state-of-the-art' computer at the University of Manchester. During the shear stage it took 15 months to achieve a deviator strain of 7 %. Research was mainly motivated by soil mechanics involving 2D quasi-static simulations of discs with linear springs. However, a notable exception was Walton (1983, 1984) whose motivation was granular flow problems. An example is illustrated in Fig. 1.3 which shows the transfer of material from one inclined surface to another sloping in the opposite direction. The figure is in fact a snapshot taken from a computer generated movie showing the particles on the left and the corresponding particle velocities on the right. It is notable that the particles are not discs but irregular polygons.

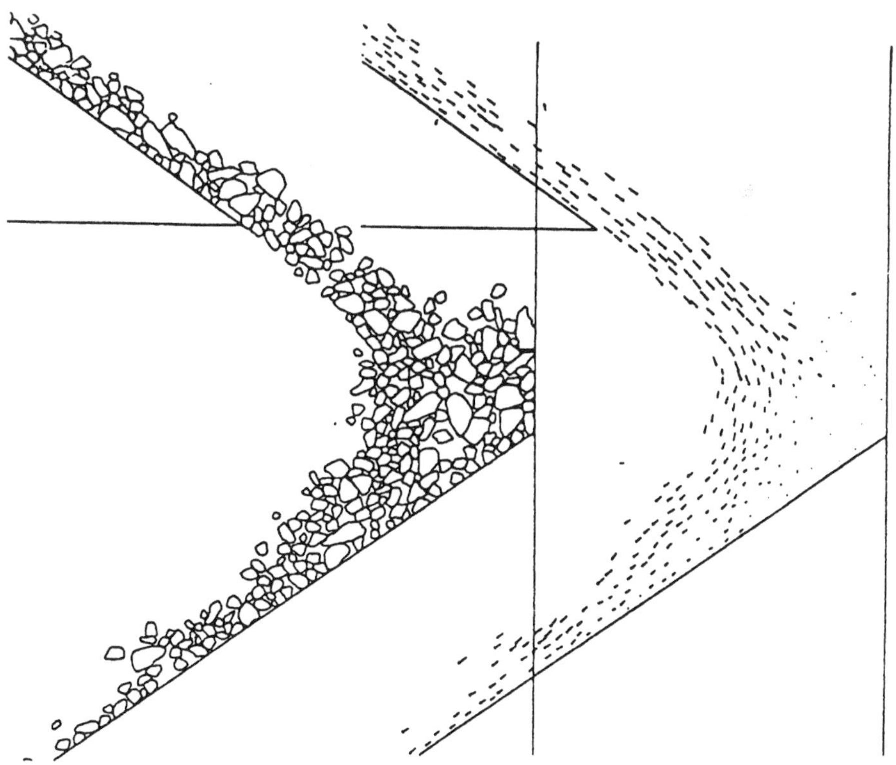

Fig. 1.3 Flow of crushed oil shale rubble (Walton 1984, Fig. 1)

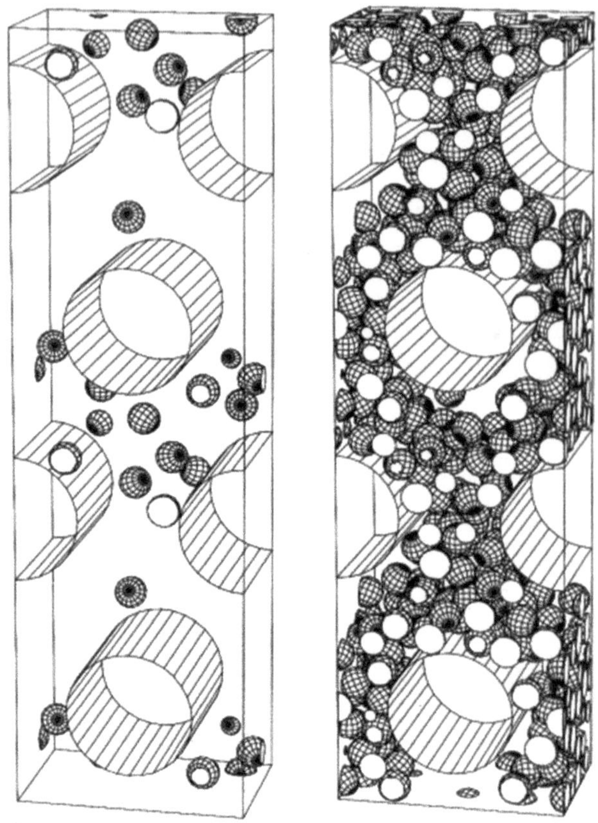

Fig. 1.4 Gravity flow through a fixed array of horizontal cylinders (Walton et al. 1988, Fig. 2)

In 1987, at a meeting in Sendai, Walton et al. (1988) presented the first 3D simulations of particle flow, as illustrated in Fig. 1.4. The figure shows, for two different particle concentrations, spheres falling under gravity through a regular array of horizontal cylinders. Note that the vertical boundaries are periodic and therefore a white circle indicates a sphere that is partially over the boundary and the other part of the sphere can be seen at the opposite boundary. At the same meeting, Cundall (1988) presented the first 3D simulations of quasi-static deformation using the TRUBAL code, as shown in Fig. 1.5. The figure shows a polydisperse system of 432 spheres in a periodic cell and the effect of the number of spheres used on the evolution of deviator stress and volumetric strain subjected to axisymmetric compression.

In the 1990s research activity became focussed on particle technology. There was an increasing number of 3D simulations with more complex contact force models, see Chaps. 3 and 4, and computer generated videos became popular. This became feasible as a result of the availability of dedicated workstations, which encouraged a significant increase in the number of researchers active in DEM simulations. The increasing number of DEM researchers was also a consequence

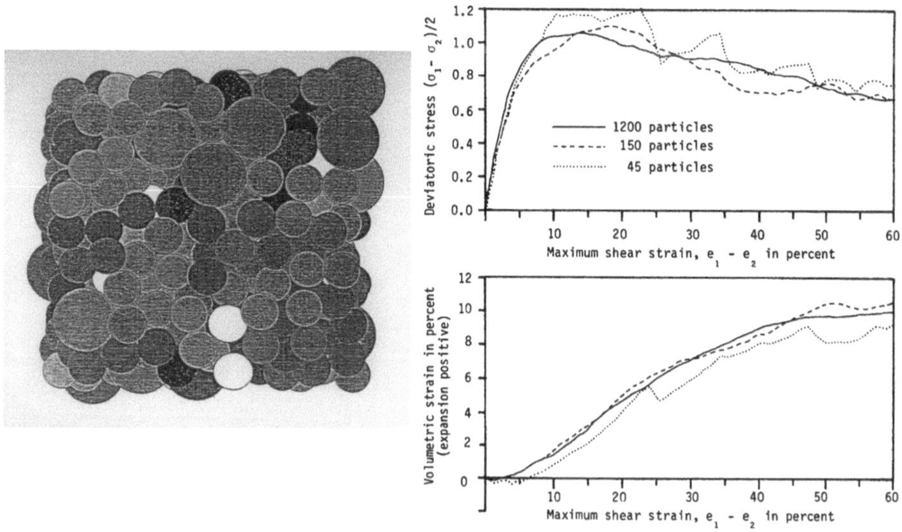

Fig. 1.5 First application of the TRUBAL code (Cundall 1988, Fig. 1)

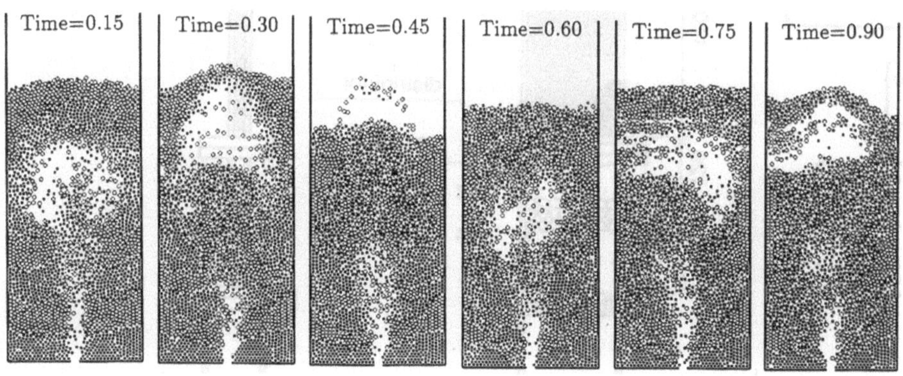

Fig. 1.6 2D DEM-CFD simulations of a gas-fluidised bed (Tsuji et al. 1993, Fig. 6d)

of the availability of commercial codes. More researchers started to consider non-spherical particle shapes. Most significant, perhaps, was the pioneering work of Tsuji et al. (1993) who combined Discrete Element Method (DEM) modelling of the particle phase with Computational Fluid Dynamics (CFD) modelling of the fluid phase to simulate gas fluidised beds, see Fig. 1.6. More details and discussion of DEM-CFD simulations will be provided in Chap. 7.

In the context of particle technology, after 1995, DEM became widely used to investigate hopper flow, rotating drums, mixers and other particle handling problems. Some examples are shown below (Figs. 1.7, 1.8, 1.9 and 1.10).

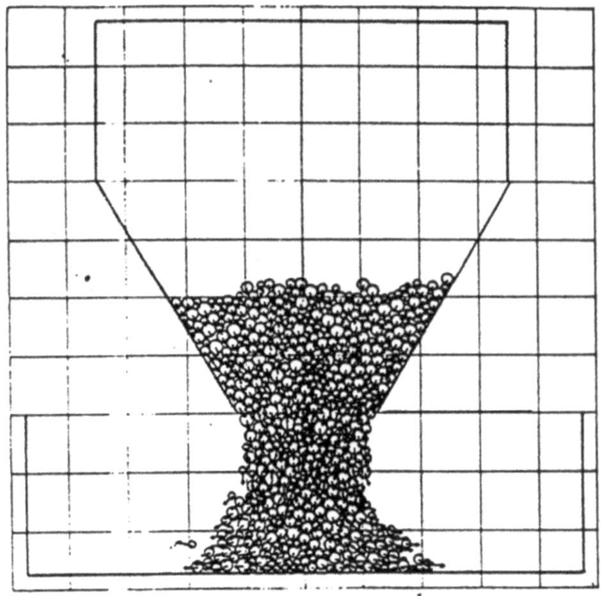

Fig. 1.7 Hopper discharge (Thornton 1991)

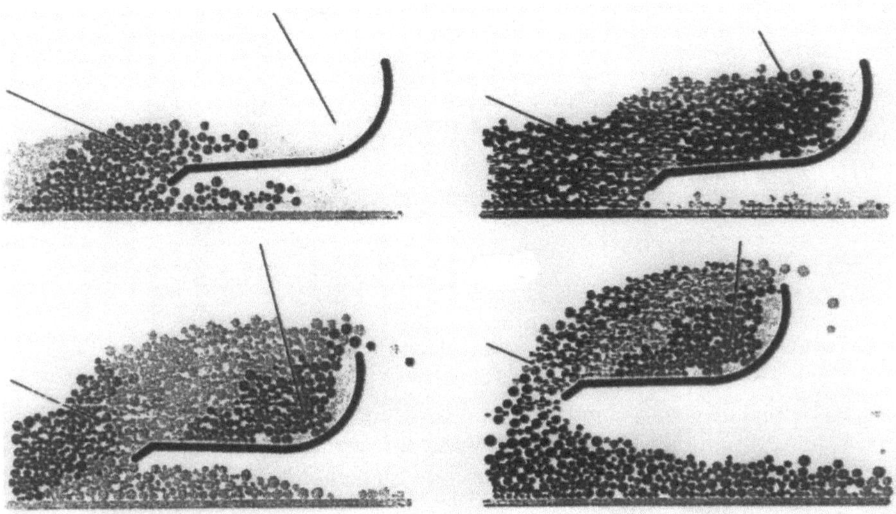

Fig. 1.8 Drag line bucket excavator (adapted from Cleary 1998, Fig. 7)

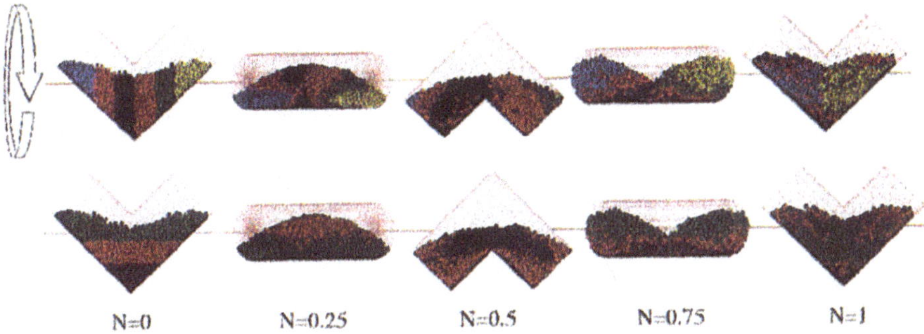

Fig. 1.9 V-mixer – N is the number of complete rotations (Moakher et al. 2000, Fig. 8)

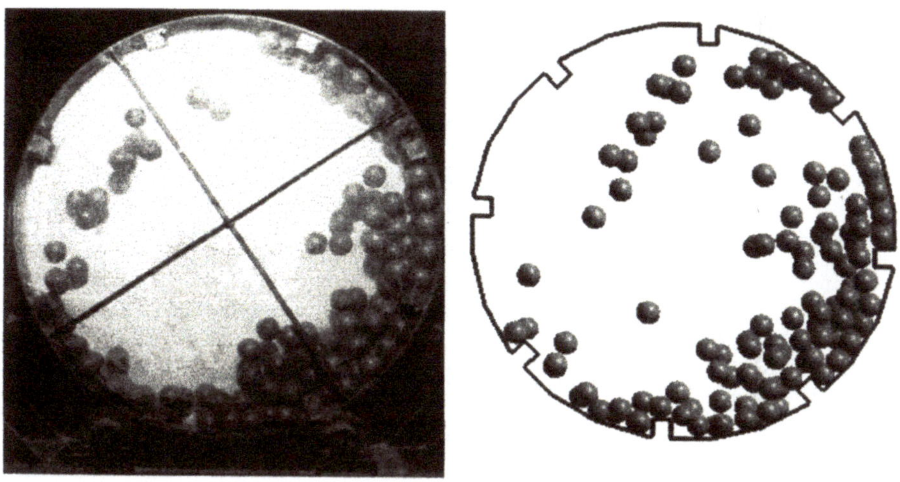

Fig. 1.10 Ball mill – experiment (*left*) simulation (*right*) (Rajamani et al. 2000, Fig. 4)

1.3 Outline of the Book

Following this introductory chapter, the second chapter provides the theoretical background to 3D DEM simulations dealing with particle kinematics and contact force calculations in general terms.[1] Chapter 3 deals with the detailed calculation of contact forces assuming that the particles are elastic, elastic-plastic or adhesive spheres. Other particle-particle interaction laws are often used by DEM researchers and these are discussed in Chap. 4. Chapters 5, 6, 7 and 8 provide details of various

[1] Throughout the book, the Einstein tensor subscript notation is used including the Kroneker delta: $\delta_{ij} = 1$ if $i = j$ and $\delta_{ij} = 0$ if $i \neq j$. Also, the soil mechanics convention that compression is positive is adopted and, hence, anticlockwise rotations are positive.

applications of DEM published by the Author. Chapter 5 deals with single particle impact and agglomerate impact simulations are presented in Chap. 6. Chapter 7 reports recent work on fluidised bed simulations. Finally in Chap. 8, simulations of quasi-static deformation of compact, polydisperse systems of spheres, the Author's original area of interest, are presented. The book is not intended to be a textbook or a manual on how to do DEM simulations. For that, the reader is referred to O'Sullivan (2011).

References

Alder, B.J., Wainwright, T.E.: Studies in molecular dynamics I. General method. J. Chem. Phys. **31**, 459–466 (1959)

Campbell, C.S., Brennen, C.E.: Computer simulation of shear flows of granular material. In: Jenkins, J.T., Satake, M. (eds.) Mechanics of Granular Materials: New Models and Constitutive Relations, pp. 313–326. Elsevier, Amsterdam (1983)

Campbell, C.S., Brennen, C.E.: Computer simulation of granular shear flows. J. Fluid Mech. **151**, 161–188 (1985)

Cleary, P.W.: The filling of dragline buckets. Math. Eng. Ind. **7**, 1–24 (1998)

Cundall, P.A: A computer model for simulating progressive, large scale movements in blocky rock systems. In: Proc. Symp. Int. Society of Rock Mechanics, Nancy, Article 8 (1971)

Cundall, P.A.: Computer simulations of dense sphere assemblies. In: Satake, M., Jenkins, J.T. (eds.) Micromechanics of Granular Materials, pp. 113–123. Elsevier, Amsterdam (1988)

Cundall, P.A., Strack, O.D.L.: A discrete numerical model for granular assemblies. Geotechnique **29**, 47–65 (1979)

Cundall, P.A., Strack, O.D.L.: Modeling of microscopic mechanisms in granular material. In: Jenkins, J.T., Satake, M. (eds.) Mechanics of Granular Materials: New Models and Constitutive Relations, pp. 137–149. Elsevier, Amsterdam (1983)

Cundall, P.A., Drescher, A., Strack, O.D.L.: Numerical experiments on granular assemblies; measurements and observations. In: Vermeer, P.A., Luger, H.J. (eds.) Deformation and Failure of Granular Materials, pp. 355–370. Balkema, Rotterdam (1982)

Metropolis, N., Rosenbluth, A.W., Rosenbluth, M.N., Teller, A.H., Teller, E.: Equations of state calculations by fast computing machines. J. Chem. Phys. **21**, 1087–1092 (1953)

Moakher, M., Shinbrot, T., Muzzio, F.J.: Experimentally validated computations of flow, mixing and segregation of non-cohesive grains in 3D tumbling blenders. Powder Technol. **109**, 58–71 (2000)

O'Sullivan, C.: Particulate Discrete Element Modelling a Geomechanics Perspective. Spon Press, Abingdon, UK (2011)

Pőschel, T., Schwager, T.: Computational Granular Dynamics. Springer, Berlin (2005)

Radjai, F., Dubois, F.: Discrete-Element Modeling of Granular Materials. Wiley, Hoboken, USA (2011)

Rajamani, R.K., Mishra, B.K., Venugopal, R., Datta, A.: Discrete element analysis of tumbling mills. Powder Technol. **109**, 105–112 (2000)

Thornton, C.: Computer simulated hopper flow. In: Proc. 4th World Congress of Chemical Engineering, Karlsruhe, Article 9.1–27 (1991)

Thornton, C., Barnes, D.J.: Computer simulated deformation of compact granular assemblies. Acta Mech. **64**, 45–61 (1986)

Tsuji, Y., Kawaguchi, T., Tanaka, T.: Discrete particle simulation of two-dimensional fluidized bed. Powder Technol. **77**, 79–87 (1993)

Walton, O.R.: Particle-dynamics calculations of shear flow. In: Jenkins, J.T., Satake, M. (eds.) Mechanics of Granular Materials: New Models and Constitutive Relations, pp. 327–338. Elsevier, Amsterdam (1983)
Walton, O.R.: Application of molecular dynamics to macroscopic particles. Int. J. Eng. Sci. **22**, 1097–1107 (1984)
Walton, O.R., Braun, R.L., Mallon, R.G., Cervelli, D.M.: Particle-dynamics calculations of gravity flow of inelastic frictional spheres. In: Satake, M., Jenkins, J.T. (eds.) Micromechanics of Granular Materials, pp. 153–161. Elsevier, Amsterdam (1988)

Chapter 2
Theoretical Background

Abstract This chapter deals with the theoretical background of granular dynamics and assembly mechanics. The complete set of equations required to perform three-dimensional DEM simulations are provided, dealing with particle kinematics, contact forces, timestep and damping. The chapter then considers various characteristics of compact particle systems with enduring contacts and provides derivations of the coordination number, fabric tensor, stress tensor and assembly modulus.

2.1 Granular Dynamics

2.1.1 Particle Kinematics

In a large system of particles, the components ($i = 1, 3$) of the translational and rotational accelerations of each particle are given by the equations

$$\frac{dv_i}{dt} = \frac{\sum F_{ci}}{m_i} + g_i \tag{2.1}$$

$$\frac{d\omega_i}{dt} = \frac{\sum F_{ti} R}{I} \tag{2.2}$$

in which F_c are the contact forces acting on the particle and $F_t R$ are the moments due to the tangential components of the contact forces, g is the acceleration due to any gravity field, v and ω are the linear velocity and angular velocity of the particle respectively and I is the moment of inertia. For a solid sphere, $I = \frac{2}{5} m R^2$.

By numerically integrating (2.1) and (2.2), using an explicit central finite difference scheme, new velocities and positions of the particles are obtained from the following equations.

$$v_i^{new} = v_i^{old} + \frac{dv_i}{dt}\Delta t \quad \text{and} \quad \omega_i^{new} = \omega_i^{old} + \frac{d\omega_i}{dt}\Delta t \tag{2.3}$$

$$x_i^{new} = x_i^{old} + v_i^{new}\Delta t \quad \text{and} \quad \theta_i^{new} = \theta_i^{old} + \omega_i^{new}\Delta t \tag{2.4}$$

where x_i and θ_i are the coordinates and components of angular rotation of the particle, Δt is the small timestep used to advance the simulation. From the new

C. Thornton, *Granular Dynamics, Contact Mechanics and Particle System Simulations*, Particle Technology Series 24, DOI 10.1007/978-3-319-18711-2_2

positions and velocities of the particles new contact forces can be calculated. There is then a need to check for new contacts and contacts that have been broken. For two spheres A and B, contact exists if the distance between their centres is less than the sum of the two radii, i.e.

$$\alpha = R_A + R_B - D \tag{2.5}$$

where α is the relative approach in the normal direction and D is the distance between the sphere centres.

If the translational and rotational velocities of the two spheres are v_i^A, ω_i^A and v_i^B, ω_i^B then the relative normal displacement increment at the contact is

$$\Delta\alpha = \left(v_i^B - v_i^A\right) n_i \Delta t \tag{2.6}$$

where n is the unit vector normal to the contact plane and directed from sphere A to sphere B. The timestep Δt will be discussed later.

The relative tangential surface displacement increment is obtained from

$$\begin{aligned} \Delta\delta_i = \left(v_i^B - v_i^A\right)\Delta t - \Delta\alpha n_i - \left(\omega_{i+1}^A n_{i+2} - \omega_{i+2}^A n_{i+1}\right) R_A \Delta t \\ - \left(\omega_{i+1}^B n_{i+2} - \omega_{i+2}^B n_{i+1}\right) R_B \Delta t \end{aligned} \tag{2.7}$$

in which the subscripts i, $i+1$, $i+2$ are rotated between the limits 1 to 3, i.e. if $i = 2$, $i+2 = 4 - 3 = 1$.

2.1.2 Contact Forces

Having obtained the relative incremental displacements at a contact, the new contact reactions can be calculated. The normal force can be updated incrementally using

$$F_n^{new} = F_n^{old} + k_n \Delta\alpha \tag{2.8}$$

However, it is better to use the functional form, e.g. $F_n = k_n \alpha$, using double precision arithmetic, if the contact force model permits it, see Chap. 3 for details. The tangential force is much more complicated and, necessarily, has to be updated incrementally, as explained below.

During particle-particle interactions the contact plane continuously rotates and it is, therefore, necessary to reorientate the current tangential force and tangential displacement directions to be orthogonal to the new contact normal direction before the tangential forces and displacements are updated. The rotation of the contact plane is given by

$$\Omega_i = \left(\left(v_{i+2}^B - v_{i+2}^A\right)\Delta t - \Delta\alpha n_{i+2}\right)\frac{n_{i+1}}{D} - \left(\left(v_{i+1}^B - v_{i+1}^A\right)\Delta t - \Delta\alpha n_{i+1}\right)\frac{n_{i+2}}{D} \quad (2.9)$$

where D is the distance between the centres of the two spheres. The directions of the tangential force and displacement are adjusted using

$$F_{ti} = \Omega_{i+1} F_{t(i+2)}^{old} - \Omega_{i+2} F_{t(i+1)}^{old} \quad \text{and} \quad \delta_i = \Omega_{i+1}\delta_{t(i+2)}^{old} - \Omega_{i+2}\delta_{t(i+1)}^{old} \quad (2.10)$$

where F_{ti}^{old} and δ_i^{old} are the components of the tangential force and tangential displacement before rotation of the contact plane and the adjusted tangential force and displacement are obtained from

$$F_t = (F_{ti}F_{ti})^{1/2} \quad \text{and} \quad \delta = (\delta_i\delta_i)^{1/2} \quad (2.11)$$

Having accounted for the rotation of the contact normal vector, a further complication arises in 3D simulations. Generally, the direction of the tangential displacement increment, defined by Eq. (2.7), is not coaxial with the direction of the old tangential displacement, defined by Eq. (2.10), as shown in Fig. 2.1a. The new tangential displacement is obtained by vector summation, i.e.

$$\delta_i^{new} = \delta_i^{old} + \Delta\delta_i \quad (2.12)$$

However, in almost all codes, the new tangential force is calculated as $F_t^{new} = F_t^{old} + k_t\Delta\delta$ and this is not exactly correct. (This can be demonstrated by considering a constant normal force, i.e. the contact area does not change. In this case, it will be found that, due to the non-coaxiality, the tangential force reaches the limiting value of $F_t = \mu F_n$ at the wrong value of δ.)

For clarity, in Fig. 2.1, the magnitude of the tangential displacement increment has been grossly exaggerated relative to the magnitude of the old tangential

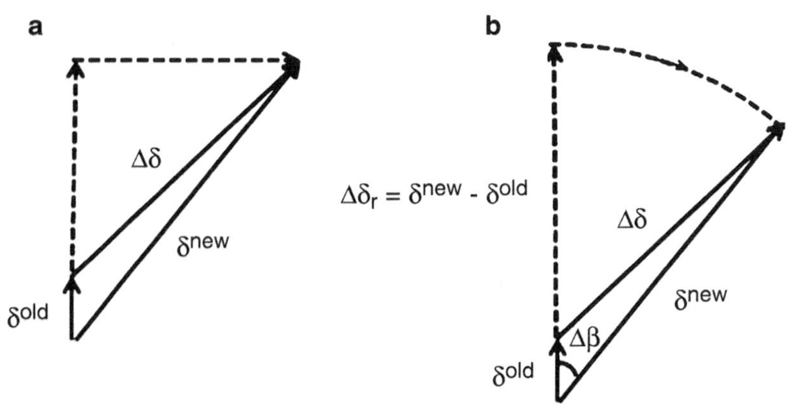

Fig. 2.1 Non-coaxiality on the contact plane considering (**a**) Cartesian and (**b**) cylindrical local coordinates

displacement. From Fig. 2.1a, it is tempting to consider that $\Delta\delta$ has two components, one that is coaxial with δ^{old} and one that is orthogonal to the direction of δ^{old}. If this was the case it would raise the question as to what is the reaction to this orthogonal component of $\Delta\delta$.

The correct solution is obtained by considering cylindrical coordinates, as shown in Fig. 2.1b. There is a normal displacement that creates a normal reaction force, a radial tangential displacement that creates a tangential reaction force and a rotational displacement (a twist) that is reacted to by a contact moment. Consequently, the tangential force is updated using

$$F_t^{new} = F_t^{old} + k_t \Delta\delta_r \tag{2.13}$$

using the scalar product

$$\Delta\delta_r = \delta^{new} - \delta^{old} \tag{2.14}$$

and the tangential force components are calculated from

$$F_{ti}^{new} = F_t^{new} \frac{\delta_i^{new}}{|\delta^{new}|} \tag{2.15}$$

The updated resultant tangential force is compared with the sliding criterion. If the sliding criterion is exceeded then the tangential force is reset to the limiting value $F_t = \mu F_n$.

Figure 2.1b also shows that, as a consequence of the non-coaxiality, there is a rotational displacement increment indicated by the angle $\Delta\beta$. However, due to the small contact area, the resultant contact moment is generally ignored.

In the above, the contact stiffnesses k_n and k_t are the current stiffnesses and depend on the contact force model used, which will be discussed in Chaps. 3 and 4.

2.1.3 Timestep

During its movement, a particle may collide with neighbouring particles or walls. However, the particle movement is also affected by particles far beyond its local neighbourhood through the propagation of disturbance waves. This problem is solved by selecting a suitably small value for the timestep such that, during a single timestep, a disturbance can only propagate from a particle to other particles in contact with it. For the case when linear springs are used to model the contact stiffnesses the critical timestep is related to the ratio of the contact spring stiffness to the particle density. For non-linear springs (e.g. a Hertzian spring) the critical timestep cannot be calculated a priori. However, it was shown by Miller and Pursey (1955) that the Rayleigh waves account for 67 % of the radiated energy in comparison with the dilational (7 %) and distortional (26 %) waves. In the

Birmingham DEM code, it is assumed that all of the energy is transferred by Rayleigh waves. This is a good approximation since the difference between the Rayleigh wave speed and the distortional wave speed is very small and the energy transferred by the dilational wave is negligible. In addition, the average time of arrival of the Rayleigh wave at any contact is the same irrespective of the location of the contact point. Consequently, the critical timestep is based on the minimum particle size and a fraction of the critical timestep is used in the simulations. The critical timestep is given by the following equation.

$$\Delta t_c = \frac{\pi R_{min}}{v_R} = \frac{\pi R_{min}}{\lambda}\sqrt{\frac{\rho}{G}} \tag{2.16}$$

where R_{min} is the minimum particle radius, ρ is the particle density, G is the particle shear modulus, v_R is the Rayleigh wave speed and λ can be obtained from

$$\left(2-\lambda^2\right)^4 = 16\left(1-\lambda^2\right)\left[1-\lambda^2\left(\frac{1-2\nu}{2(1-\nu)}\right)\right] \tag{2.17}$$

which can be approximated by

$$\lambda = 0.8766 + 0.1631\nu \tag{2.18}$$

where ν is the Poisson's ratio of the particle.

2.1.4 Damping

Unlike most other DEM codes, the Birmingham code does not include a dashpot force as part of the contact force. There are, however, dashpots that are used to dissipate a small amount of energy due to elastic wave propagation through a solid particle. The dashpot forces are not considered to be part of the actual contact forces but they are added to the normal and tangential contact forces to provide the contribution to the out-of-balance force and moment acting on the particle from that contact, used in Eqs. (2.1) and (2.2). The dashpot forces are given by

$$F_{nd} = 2\beta\sqrt{m^* k_n}\Delta\alpha/\Delta t \quad \text{and} \quad F_{td} = 2\beta\sqrt{m^* k_t}\Delta\delta_r/\Delta t \tag{2.19}$$

$$\text{with} \quad \frac{1}{m^*} = \frac{1}{m_A} + \frac{1}{m_B} \tag{2.20}$$

where m_A and m_B are the masses of the two particles and the value of β in Eq. (2.19) is in the range 0.01–0.03.

2.2 Assembly Mechanics

When considering compact systems of particles with enduring contacts there is a need to define certain characteristics such as stability, microstructure and the state of stress for the system.

2.2.1 Stability Considerations

The structural stability of a system of particles is related to the average number of contacts per particle, i.e. the coordination number $Z = 2C/N$ where C is the number of contacts, N is the number of particles and the 2 accounts for the fact that each contact is shared by two particles.

Consider a 3D system of particles with $\mu = \infty$, i.e. no sliding occurs at any contact. The number of degrees of freedom of a single particle is 6 (3 rotations and 3 translations) and the total number of degrees of freedom in the system is $6N$. The number of constraints (unknown reactions) at a single contact is 3 (a normal force, a radial tangential force and a contact moment) and the total number of constraints in the system is $3C$. If the total number of constraints equals the total number of degrees of freedom, i.e. $3C = 6N$, the system is statically determinate (isostatic) and we can define a critical coordination number $Z_c = 4$. If $Z > Z_c$ then the system is indeterminate (hypostatic) which means that there are more contacts than necessary to ensure stability; it is a redundant system. If $Z < Z_c$ then equilibrium cannot be satisfied since there are fewer contacts than necessary to keep the system stable and the system becomes a mechanism (hyperstatic).

Defining the coordination number as $Z = 2C/N$ is not necessarily appropriate since this is the apparent coordination number and includes particles with no contacts. We can define a geometrical coordination number

$$Z_g = \frac{2C}{(N - N_0)} \tag{2.21}$$

where N_0 is the number of particles with no contacts. This geometrical coordination number can be useful to characterise clusters or agglomerates. However, this definition includes particles with only one contact and these do not contribute to the stability of the system. Therefore, in order to examine stability, we define a mechanical coordination number

$$Z_m = \frac{(2C - N_1)}{(N - N_1 - N_0)} \tag{2.22}$$

where N_1 is the number of particles with only one contact. Consequently we define a system to be isostatic when $Z_m = 4$.

The above is a limiting case. The other limiting case is when $\mu = 0$. In this case there are no particle rotations and the only constraint at a contact is the normal force. Therefore, for $\mu = 0$, the isostatic state corresponds to $Z_m = 6$. Both in reality and in simulations the interparticle friction is finite and, consequently, the critical value of Z_m depends on the percentage of sliding contacts and it is difficult to ascertain exactly what the value should be, but $Z_m = 4$ can be taken as a lower bound.

An alternative is to define a redundancy index I_R that is a function of the fraction of contacts that are sliding. The redundancy index is defined as the ratio of the number of constraints (allowing for the reduced number of constraints where sliding occurs) to the number of degrees of freedom in the system. Therefore in 3D, if the contact moment referred to in Sect. 2.1.2 is ignored, then we can write

$$I_R = \frac{C}{N}\left(\frac{3 - 2f}{6}\right) \tag{2.23}$$

where f is the fraction of sliding contacts, e.g. 0.10. If $I_R = 1$ the system is isostatic, hypostatic if $I_R > 1$ and hyperstatic if $I_R < 1$. If the particles with no contacts are ignored and the number of particles with only one contact is negligibly small then Eq. (2.23) can be rewritten as

$$I_R = Z_m\left(\frac{3 - 2f}{12}\right) \tag{2.24}$$

2.2.2 *Microstructure*

It is universally recognised that the mechanical behaviour of granular material is strongly influenced by the closeness of the packing of the individual particles. Consequently, it is common practice to characterise granular material by a scalar parameter, e.g. void ratio, porosity or solid fraction. However, it is now well known that the microstructure of granular materials is anisotropic and this requires a tensorial description. Satake (1982) suggested that, for disc or sphere systems, the structural anisotropy is defined by the orientational distribution of contact normals n_i that can be characterised by a second order fabric tensor ϕ_{ij}

$$\phi_{ij} = \frac{1}{2C}\sum_{1}^{2C} n_i n_j = \langle n_i n_j \rangle \tag{2.25}$$

where C is the total number of contacts, which are counted twice since each contact belongs to two particles. For discs and sphere assemblies, the direction cosines of a unit contact normal vector are given by

$$n_i = \left(x_i^A - x_i^B\right) / \left(R^A + R^B\right) \tag{2.26}$$

where x_i^A and x_i^B are the positions of the centres of the two contacting particles of radii R^A and R^B. Although higher order fabric tensors such as $\phi_{ijkl} = \langle n_i n_j n_k n_l \rangle$ may also be considered DEM simulations have shown that a second order representation, as given by (2.25), is satisfactory

If we consider a continuous distribution of contact normal vectors we can also write

$$\phi_{ij} = \int_\Omega E(n) n_i n_j d\Omega \tag{2.27}$$

which satisfies the conditions

$$\int_\Omega E(n)\, d\Omega = 1 \quad \text{and} \quad E(n) = E(-n) \tag{2.28}$$

where $E(n)$ is the probability density function of contact normals and $E(n)d\Omega$ gives the estimated rate of n whose directions are within a small solid angle $d\Omega$. The probability density function can be expressed as a Fourier series which, in tensorial form, can be written as

$$E(n) = E_0 + E_{ij} f_{ij} \tag{2.29}$$

where

$$f_{ij} = n_i n_j - \frac{\delta_{ij}}{3} \tag{2.30}$$

The Fourier coefficients are given by

$$E_0 = \frac{1}{4\pi} \int_\Omega E(n) d\Omega = \frac{1}{4\pi} \tag{2.31}$$

$$E_{ij} = \frac{15}{8\pi} \int_\Omega E(n) f_{ij} d\Omega = \frac{15}{8\pi} \langle f_{ij} \rangle = \frac{15}{8\pi} \left(\phi_{ij} - \frac{\delta_{ij}}{3} \right) \tag{2.32}$$

These coefficients are traceless, completely symmetric and transform as tensors under rigid body rotations. In 2D systems we can write

$$\phi_{ij} = \int_0^{2\pi} E(\theta)\, n_i n_j d\theta \tag{2.33}$$

where θ is the inclination angle of n with respect to the reference axis and

$$E(\theta) = E_0 + E_{ij} f_{ij} \tag{2.34}$$

$$\text{with } E_0 = \frac{1}{2\pi} \int_0^{2\pi} E(\theta) d\theta = \frac{1}{2\pi} \quad \text{and} \quad E_{ij} = \frac{2}{\pi} \int_0^{2\pi} E(\theta) f_{ij} d\theta = \frac{2}{\pi} \left\langle f_{ij} \right\rangle \tag{2.35}$$

Alternatively $E(\theta)$can be expressed as

$$E(\theta) = a_0 + a_2 \cos(2\theta) + b_2 \sin(2\theta) \tag{2.36}$$

with

$$a_0 = \frac{1}{2\pi}; \; a_2 = \frac{2}{\pi} \langle f_{11} \rangle = \frac{2}{\pi} \left(\phi_{11} - \frac{1}{2} \right); \; b_2 = \frac{2}{\pi} \langle f_{12} \rangle = \frac{2}{\pi} \phi_{12} \tag{2.37}$$

It is of interest to note that if we assume a second order Fourier series representation then this implies limits on the degree of anisotropy that in 2D can be defined by the deviator fabric $(\phi_1 - \phi_2)$. If the principal axes of anisotropy coincide with the reference axes then

$$E(\theta) = \frac{1}{2\pi} + a_2 \cos(2\theta) \tag{2.38}$$

Since $E(\theta)$ cannot be negative we take $E(\theta) = 0$ at $\theta = \pi/2$ to be the limit condition and, using Eq. 2.37, we obtain $\phi_1 = 0.75$ and $\phi_2 = 0.25$. Consequently, the maximum possible deviator fabric is 0.5. A similar exercise in 3D yields a corresponding maximum possible deviator fabric of 0.25, in the case of axisymmetric compression.

2.2.3 Stress

Particle systems sometimes behave like a solid and sometimes like a liquid. In general, the state of stress is defined by the following equation.

$$\sigma_{ij} = p\delta_{ij} + \frac{1}{2V} \sum^{N_p} m \tilde{\dot{x}}_i \tilde{\dot{x}}_j + \frac{1}{V} \int \sigma_{ij}^p dV \tag{2.39}$$

where V is the volume of the system containing N_p particles, m is the mass of a particle with a fluctuating velocity $\tilde{\dot{x}} = \dot{x} - \langle \dot{x} \rangle$ and σ_{ij}^p is the average state of stress inside a particle.

The first term on the right hand side is due to the fluid pressure p which contributes to the isotropic component of the stress tensor. The second term is the fluctuating kinetic energy density, or Reynolds stress, which is related to the so-called 'granular temperature' and is, in general, anisotropic. The third term is the Cauchy stress due to the particle-particle interactions, either collisional or enduring.

The fluctuating velocities arise from the particle collision forces and the drag forces, lift forces, etc., or anything else that causes the particles to deviate from the fluid streamlines. In quasi-static deformation of compact systems, when particle interactions are enduring, the second term is negligibly small and we can define, what is known in soil mechanics as, the effective stress, σ'_{ij}

$$\sigma'_{ij} = \sigma_{ij} - p\delta_{ij} = \frac{1}{V}\int \sigma^{p}_{ij}\, dV \tag{2.40}$$

The average stress tensor for a single particle may be written as

$$\sigma^{p}_{ij} = \frac{1}{V^{p}}\int \sigma_{ij}\, dV^{p} \tag{2.41}$$

where V^p is the volume occupied by a single particle. (Note that the volume occupied by the particle V^p is not the solid volume of the particle but also includes the portion of the void volume that is more close to the particle than to any other particle.) The volume integral can be replaced by a surface integral, using the Divergence Theorem, to give

$$\sigma^{p}_{ij} = \frac{1}{V^{p}}\int x_i t_j\, dS \tag{2.42}$$

By considering the tractions t_j to consist of discrete forces F_j acting at point contacts defined by the coordinates x_i the integral in Eq. (2.42) may be replaced by a summation over the n contacts of particle p. Thus

$$\sigma^{p}_{ij} = \frac{1}{V^{p}}\sum_{1}^{n} x_i F_j \tag{2.43}$$

Since the effective stress is not continuously distributed over the system ($\sigma'_{ij} = 0$ in the voids) Eq. (2.40) may be rewritten as a summation

$$\sigma'_{ij} = \frac{1}{V}\sum^{N_p} \sigma^{p}_{ij} V^{p} = \frac{1}{V}\sum^{N_p}\sum^{n} x_i F_j \tag{2.44}$$

noting that for each particle the coordinates x_i are referenced to the particle centre and, for systems of discs or spheres, $x_i = Rn_i$. Also, the contact force F may be partitioned into the normal and tangential components, F_n and F_t, so that $F_{ni} = F_n n_i$ and $F_{ti} = F_t t_i$, where n_i define the contact normal vector and t_i is orthogonal to n_i.

For polydisperse systems of spheres (or discs) the summation is simply carried out over the C contacts in the system to give

$$\sigma'_{ij} = \frac{1}{V}\sum^{C}\left(R^A + R^B\right)F_n n_i n_j + \frac{1}{V}\sum^{C}\left(R^A + R^B\right)F_t n_i t_j \tag{2.45}$$

or, in terms of statistical averages, we may write

$$\sigma'_{ij} = \frac{C}{V}\langle DF_n n_i n_j\rangle + \frac{C}{V}\langle DF_t n_i t_j\rangle \tag{2.46}$$

where $D = R^A + R^B$ is the distance between the centres of the two contacting spheres.

2.2.4 Assembly Modulus

Following Walton (1987) we may construct an equation relating a small change in the ensemble average stress to small changes in the interparticle contact forces. Then, by making a simplifying assumption about the applied strain increment field we may obtain the current effective modulus/compliance of an assembly of particles. In other words, we wish to establish the following relationships

$$d\sigma_{ij} = S_{ijkl}d\varepsilon_{kl} \quad \text{or} \quad d\varepsilon_{ij} = C_{ijkl}d\sigma_{kl} \quad \text{with} \quad C_{ijkl} = \left(S_{ijkl}\right)^{-1} \tag{2.47}$$

Taking the pore fluid pressure to be zero, Eq. (2.46) may be written as

$$\sigma_{ij} = \frac{C}{V}\langle Dn_i F_j\rangle \tag{2.48}$$

or, in incremental form

$$d\sigma_{ij} = \frac{C}{V}\langle Dn_i\, dF_j\rangle \tag{2.49}$$

If k_n and k_t are the current normal and tangential contact stiffnesses then

$$dF_i = k_n n_i d\alpha + k_t d\delta_{ri} \tag{2.50}$$

where $d\alpha$ is the increment in the relative approach of the two particles and $d\delta_r$ is the relative radial tangential displacement increment at the contact. It follows that

$$d\sigma_{ij} = \frac{C}{V}\left[\left\langle Dk_n d\alpha n_i n_j \right\rangle + \left\langle Dk_t d\delta_{ri} n_j \right\rangle\right] \tag{2.51}$$

To proceed further, we assume that the incremental displacement du_i of the sphere centres is consistent with an applied uniform strain increment field, i.e.

$$du_i = d\varepsilon_{ij} X_j \tag{2.52}$$

where X_{ij} defines the coordinates of a sphere and $d\varepsilon_{ij}$ is the applied strain increment tensor.

In a polydisperse system of spheres the individual sphere rotations significantly affect the relative tangential displacement increment at the contact, see Eq. (2.7), in a random way and are independent of the applied strain increment tensor. They are therefore difficult to incorporate into the macroscopic stress-strain description. Consequently if, for simplicity, we assume no particle rotations then $d\delta_{ri} = d\delta_i$ and the normal and tangential relative incremental displacements at the contact are

$$d\alpha = \left(du_i^B - du_i^A\right) n_i \tag{2.53}$$

$$d\delta_i = \left(du_i^B - du_i^A\right) - d\alpha\, n_i \tag{2.54}$$

However

$$\left(du_i^B - du_i^A\right) = d\varepsilon_{ij}\left(X_j^B - X_j^A\right) = d\varepsilon_{ij} D n_j \tag{2.55}$$

Therefore

$$d\alpha = D d\varepsilon_{ij} n_j n_i \tag{2.56}$$

$$d\delta_i = D d\varepsilon_{ij} n_j - D d\varepsilon_{kl} n_k n_l n_i \tag{2.57}$$

which may be substituted into Eq. 2.51 to give

$$\begin{aligned} d\sigma_{ij} = \frac{C}{V}\Big[&\left\langle k_n D^2 d\varepsilon_{ki} n_i n_j n_l n_k \right\rangle + \left\langle k_t D^2 d\varepsilon_{jk} n_i n_k \right\rangle \\ &- \left\langle k_t D^2 d\varepsilon_{kl} n_i n_l n_k n_j \right\rangle\Big] \end{aligned} \tag{2.58}$$

Noting that the second term is not necessarily symmetric we write

$$\begin{aligned} d\sigma_{ij} = \frac{C}{V}\Big[&\left\langle k_n D^2 d\varepsilon_{kl} n_k n_l n_i n_j \right\rangle - \left\langle k_t D^2 d\varepsilon_{kl} n_k n_l n_i n_j \right\rangle \\ &+ 0.5\left(\left\langle k_t D^2 d\varepsilon_{ik} n_k n_j \right\rangle + \left\langle k_t D^2 d\varepsilon_{jk} n_k n_i \right\rangle\right)\Big] \end{aligned} \tag{2.59}$$

Using $d\varepsilon_{ik} = d\varepsilon_{kl}\delta_{il}$ and $d\varepsilon_{jk} = d\varepsilon_{kl}\delta_{jl}$ we obtain $d\sigma_{ij} = S_{ijkl} d\varepsilon_{kl}$ with

$$S_{ijkl} = \frac{C}{V}\left[\left\langle k_n D^2 n_i n_j n_k n_l \right\rangle - \left\langle k_t D^2 n_i n_j n_k n_l \right\rangle \right.$$
$$+ 0.25\left(\left\langle k_t D^2 n_j n_k \right\rangle \delta_{il} + \left\langle k_t D^2 n_i n_k \right\rangle \delta_{jl}\right.$$
$$\left.\left. + \left\langle k_t D^2 n_j n_i \right\rangle \delta_{ik} + \left\langle k_t D^2 n_i n_l \right\rangle \delta_{jk}\right)\right] \quad (2.60)$$

An example of the use of Eq. (2.60) will be provided in Chap. 8.

References

Miller, G.F., Pursey, H.: On the partition of energy between elastic waves in a semi-infinite solid. Proc. R. Soc. A **233**, 55–69 (1955)

Satake, M.: Fabric tensor in granular materials. In: Vermeer, P.A., Luger, H.J. (eds.) Deformation and Failure of Granular Materials, pp. 63–68. Balkema, Rotterdam (1982)

Walton, K.: The effective elastic moduli of a random packing of spheres. J. Mech. Phys. Solids **35**, 213–226 (1987)

Chapter 3
Contact Mechanics

Abstract As a result of the relative displacements between contiguous particles, forces are generated at the interparticle contacts. The contact force-displacement behaviour depends on the material properties of the particles, the sizes of the two particles in contact and the surface conditions. In this chapter we consider contact force models that are based on theoretical contact mechanics. Equations are derived for both the normal and tangential contact forces between two spheres in contact for elastic and elastic-plastic interactions, with and without adhesion.

The application of external strain fields to granular materials causes relative movement between the constitutive particles. As a result of the relative displacements between contiguous particles, forces are generated at the interparticle contacts and the macroscopic state of stress is a function of the distribution of contact forces within the material. The contact force-displacement behaviour depends on the material properties of the particles, the sizes of the two particles in contact and the surface conditions. In this chapter we restrict our consideration to spheres although the theories can be generalised to other shapes provided that the surfaces are non-conforming and the curvature of the two contacting particles at the point of contact is known (Johnson 1985).

3.1 Elastic Interactions

The normal and tangential force-displacement relationships for elastic spheres with friction are provided by the theories of Hertz (1881), Mindlin (1949), and Mindlin and Deresiewicz (1953). The normal and tangential contact forces are obtained by integrating the respective normal and tangential traction distributions over the contact area.

C. Thornton, *Granular Dynamics, Contact Mechanics and Particle System Simulations*, Particle Technology Series 24, DOI 10.1007/978-3-319-18711-2_3

3.1.1 Normal Interaction

For two contacting elastic spheres of radii R_i ($i = 1, 2$), the theory of Hertz (1881) predicts a semi-ellipsoidal normal pressure distribution acting over the small circular contact area of radius a, and expressed as

$$p(r) = p_0\left[1 - \left(\frac{r}{a}\right)^2\right]^{1/2} \tag{3.1}$$

which leads to normal displacements over the contact area as illustrated in the cross-section shown in Fig. 3.1 and given by

$$u_i(r) = \frac{\pi\, p_0\left(1 - \nu_i^2\right)}{4aE_i}\left(2a^2 - r^2\right) \tag{3.2}$$

where E_i and ν_i are the Young's modulus and Poisson's ratio for the respective sphere and Eq. (3.2) satisfies the following boundary condition for the surface of the contact area

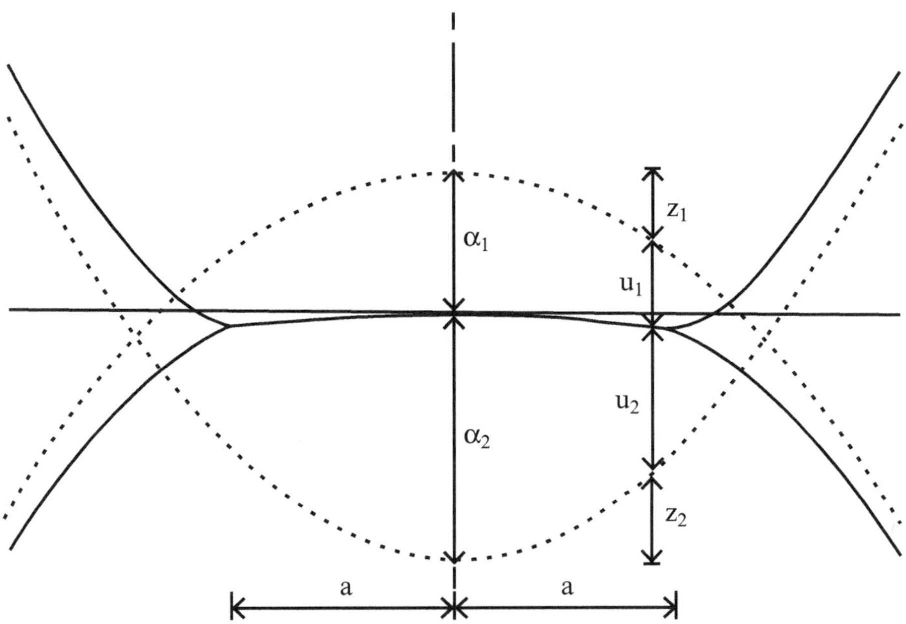

Fig. 3.1 Geometry of the local deformation at the contact between two dissimilar, unequal sized elastic spheres subjected to an applied normal force (vertical scale exaggerated)

$$u_1(r) + u_2(r) = \alpha - \left(\frac{r^2}{2R^*}\right) \tag{3.3}$$

with the relative curvature of the surface defined as

$$\frac{1}{R^*} = \frac{1}{R_1} + \frac{1}{R_2} \tag{3.4}$$

and $\alpha = \alpha_1 + \alpha_2$ is the relative approach of the centroids of the two spheres in contact. Substitution of Eq. (3.2) into Eq. (3.3) for $i = 1, 2$ leads to

$$\left(\frac{\pi p_0}{4aE^*}\right)(2a^2 - r^2) = \alpha - \left(\frac{r^2}{2R^*}\right) \tag{3.5}$$

where the relative contact compliance is

$$\frac{1}{E^*} = \frac{(1 - \nu_1^2)}{E_1} + \frac{(1 - \nu_2^2)}{E_2} \tag{3.6}$$

Substituting $r = 0$ into Eq. (3.5), the relative approach is

$$\alpha = \frac{\pi p_0 a}{2E^*} \tag{3.7}$$

and the radius of the contact area is obtained from Eqs. (3.5) and (3.7) with $r = a$ to give

$$a = \frac{\pi p_0 R^*}{2E^*} \tag{3.8}$$

The total normal force is defined as

$$F_n = \int_0^a p(r) 2\pi r dr = \frac{2}{3} p_0 \pi a^2 \tag{3.9}$$

which may be substituted into Eqs. (3.7) and (3.8) to give

$$a^3 = \frac{3F_n R^*}{4E^*} \tag{3.10}$$

and

$$\alpha^3 = \frac{9F_n^2}{16R^* E^{*2}} \tag{3.11}$$

noting also that

$$a^2 = R^* \alpha \tag{3.12}$$

Rearranging Eq. (3.11)

$$F_n = \frac{4}{3} E^* \left(R^* \alpha^3 \right)^{1/2} \tag{3.13}$$

from which the normal contact stiffness is defined as

$$k_n = \frac{dF_n}{d\alpha} = 2E^* (R^* \alpha)^{1/2} = 2E^* a \tag{3.14}$$

3.1.2 Tangential Interaction

Although it is only strictly valid for two spheres with the same elastic properties, for simplicity, we assume that the tangential interaction does not affect the normal interaction. Mindlin (1949) demonstrated that, for a constant normal force F_n, the effect of applying a tangential force $F_t < \mu F_n$ is to cause a small relative tangential motion, termed 'slip', over part of the contact area. Over the remaining part no such relative movement occurs and the surfaces are said to 'adhere' or 'stick'. The application of a tangential force initiates the formation of an annulus of slip at the perimeter of the contact area which, for a monotonically increasing tangential force, progresses radially inwards until, when $F_t = \mu F_n$, the 'stick' area is zero and sliding occurs with the tangential traction $q(r) = \mu p(r)$ for all $r \le a$.

The tangential force results in shear deformation adjacent to the contact surface as illustrated in Fig. 3.2, which shows the rigid displacements of the sphere centres δ_i, the tangential elastic displacements at the contact surface u_i and the slip displacements s_i. The resultant slip displacement

$$s = s_1 - s_2 = (u_1 - \delta_1) - (u_2 - \delta_2) = (u_1 - u_2) - (\delta_1 - \delta_2) \tag{3.15}$$

In the 'stick' region $s = 0$ therefore

$$u_1 - u_2 = \delta_1 - \delta_2 = \delta \tag{3.16}$$

For the limiting condition $F_t = \mu F_n$, the distribution of the tangential traction is given by

$$q(r) = \left(\frac{3\mu F_n}{2\pi a^3} \right) \left(a^2 - r^2 \right)^{1/2} \quad for \quad 0 \le r \le a \tag{3.17}$$

For $F_t < \mu F_n$, the corresponding traction distribution is obtained by superimposing a negative traction over the stick region of radius $b < a$

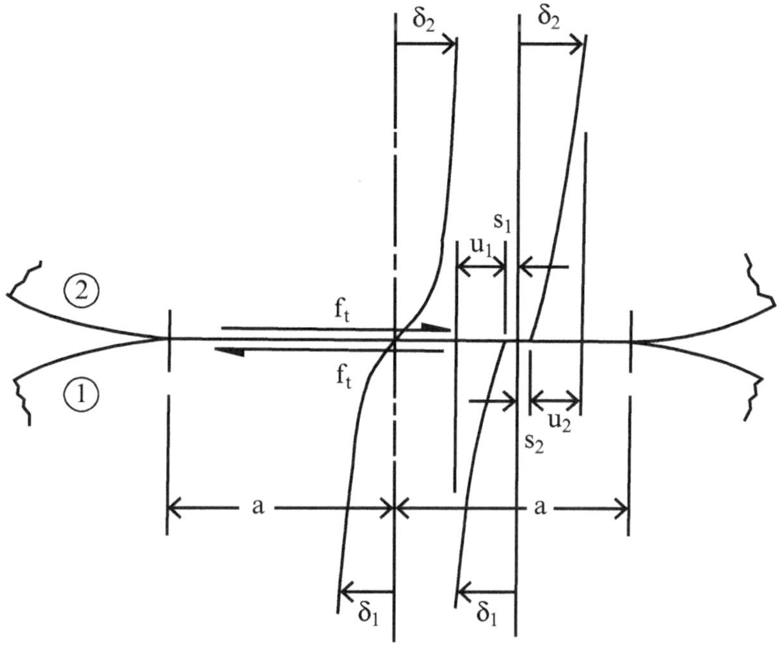

Fig. 3.2 Shear deformation adjacent to the contact due to an applied tangential force

$$q(r) = -\left(\frac{3\mu F_n}{2\pi a^3}\right)\left(b^2 - r^2\right)^{1/2} \quad for \quad 0 \le r \le b \tag{3.18}$$

Hence, the distribution of the tangential traction over the total contact area, shown in Fig. 3.3, is given as

$$q(r) = \left(\frac{3\mu F_n}{2\pi a^3}\right)\left[\left(a^2 - r^2\right)^{1/2} - \left(b^2 - r^2\right)^{1/2}\right] \quad for \quad 0 \le r \le b \tag{3.19a}$$

$$q(r) = \left(\frac{3\mu F_n}{2\pi a^3}\right)\left(a^2 - r^2\right)^{1/2} \quad for \quad b \le r \le a \tag{3.19b}$$

Mindlin (1949) showed that the relative tangential displacement of the two spheres is

$$\delta = \left(\frac{3\mu F_n}{16G^* a}\right)\left(1 - \frac{b^2}{a^2}\right) \tag{3.20}$$

where

$$\frac{1}{G^*} = \frac{(2 - \nu_1)}{G_1} + \frac{(2 - \nu_2)}{G_2} \tag{3.21}$$

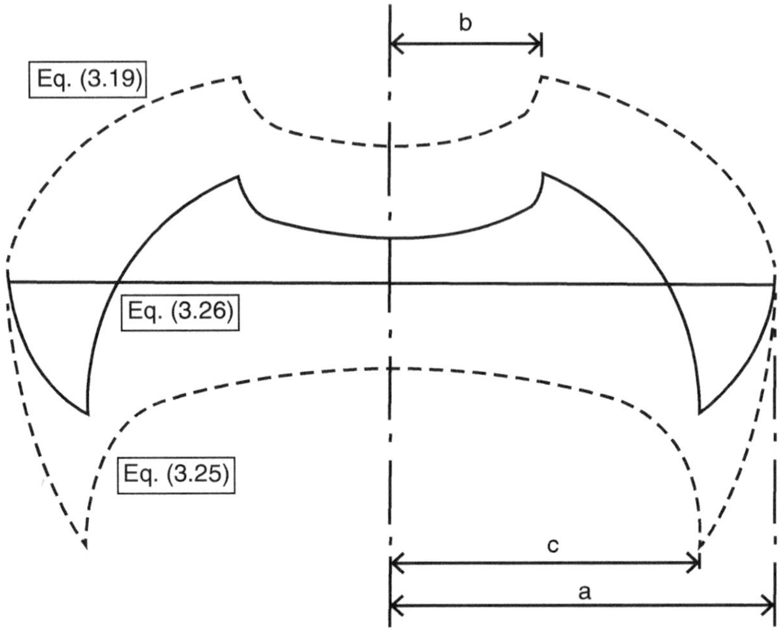

Fig. 3.3 Tangential traction distributions during loading and during unloading

The magnitude of the tangential force is defined, using Eq. (3.19), as

$$F_t = 2\pi \int_0^a q r dr = \mu F_n \left(1 - \frac{b^3}{a^3} \right) \tag{3.22}$$

Rearranging Eq. (3.22) and substituting in Eq. (3.20) we obtain the tangential force-displacement law as

$$\delta = \left(\frac{3\mu F_n}{16G^* a} \right) \left[1 - \left(1 - \frac{F_t}{\mu F_n} \right)^{2/3} \right] \tag{3.23}$$

To obtain the tangential stiffness it is convenient to differentiate Eq. (3.23) to obtain the compliance and then invert. Thus

$$k_t = \frac{dF_t}{d\delta} = 8G^* a \left(1 - \frac{F_t}{\mu F_n} \right)^{1/3} \tag{3.24}$$

The development of the slip annulus during loading involves a dissipative process and hence the slip annulus does not simply recede during unloading. If the tangential displacement is reversed then slip in the opposite direction (counterslip) spreads radially inwards from the perimeter of the contact area. The energy required to produce the annulus of counter slip is twice that needed to form the original slip annulus since the counterslip has to cancel the original slip and progress the slip in the opposite direction.

The distribution of the tangential traction at the start of unloading is given by Eq. (3.19a, b). To obtain the traction distribution during unloading, a negative traction is superimposed of the form

$$q(r) = -2\left(\frac{3\mu F_n}{2\pi a^3}\right)\left[\left(a^2 - r^2\right)^{1/2} - \left(c^2 - r^2\right)^{1/2}\right] \quad for \quad 0 \le r \le c \tag{3.25a}$$

$$q(r) = -2\left(\frac{3\mu F_n}{2\pi a^3}\right)\left(a^2 - r^2\right)^{1/2} \quad for \quad c \le r \le a \tag{3.25b}$$

where $c \le r \le a$ defines the annulus of counterslip. The resultant traction distribution is obtained by adding Eqs. (3.19a, b) and (3.25a, b) to give

$$q(r) = -\left(\frac{3\mu F_n}{2\pi a^3}\right)\left[\left(a^2 - r^2\right)^{1/2} - 2\left(c^2 - r^2\right)^{1/2} + \left(b^2 - r^2\right)^{1/2}\right] \quad for \quad 0 \le r \le b \tag{3.26a}$$

$$q(r) = -\left(\frac{3\mu F_n}{2\pi a^3}\right)\left[\left(a^2 - r^2\right)^{1/2} - 2\left(c^2 - r^2\right)^{1/2}\right] \quad for \quad b \le r \le c \tag{3.26b}$$

$$q(r) = -\left(\frac{3\mu F_n}{2\pi a^3}\right)\left(a^2 - r^2\right)^{1/2} \quad for \quad c \le r \le a \tag{3.26c}$$

which is illustrated in Fig. 3.3.

Integrating Eq. (3.26a, b, c) over the contact area to obtain the tangential force

$$F_t = \mu F_n\left(1 - \frac{b^3}{a^3}\right) - 2\mu F_n\left(1 - \frac{c^3}{a^3}\right) \tag{3.27}$$

from which

$$c = a\left[1 - \frac{\left(F_t^* - F_t\right)}{2\mu F_n}\right]^{1/3} \tag{3.28}$$

where F_t^* is the tangential force from which unloading commenced and is given by Eq. (3.22). The relative displacement of the two spheres is given by

$$\begin{aligned} \delta &= \left(\frac{3\mu F_n}{16G^* a}\right)\left(2\frac{c^2}{a^2} - \frac{b^2}{a^2} - 1\right) \\ &= \left(\frac{3\mu F_n}{16G^* a}\right)\left[2\left(1 - \frac{\left(F_t^* - F_t\right)}{2\mu F_n}\right)^{2/3} - \left(1 - \frac{F_t^*}{\mu F_n}\right)^{2/3} - 1\right] \end{aligned} \tag{3.29}$$

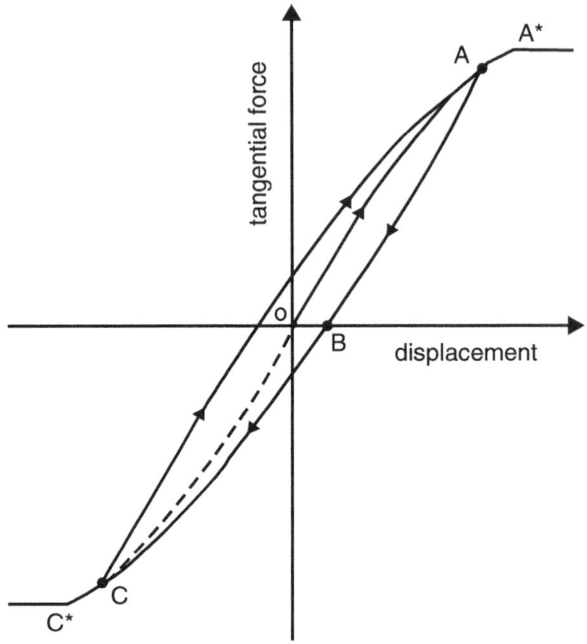

Fig. 3.4 Tangential contact force-displacement relationship

Differentiating Eq. (3.29) to obtain the compliance and then inverting provides the tangential stiffness

$$k_t = \frac{dF_t}{d\delta} = 8G^* a\left[1 - \frac{(F_t^* - F_t)}{2\mu F_n}\right]^{1/3} \tag{3.30}$$

The force-displacement curve for a reversal of the tangential displacement is shown in Fig. 3.4. If unloading commenced from the point A on the loading curve OA^*, then the tangential force decreases along ABC. The distribution of the tangential traction at point B, when $F_t = 0$, is defined by Eq. (3.26a, b, c) and illustrated in Fig. 3.3. Counterslip continues to spread radially inwards until point C is reached when $F_t = -F_t^*$ and $c = b$. Further decreases in F_t follow the inverse loading curve OCC^* in Fig. 3.4.

Following the sequence of loading and unloading, it is possible that a further reversal in the relative tangential displacement of the two spheres will occur which will result in a reloading situation. Consider that the initial tangential loading produced slip over the annulus $b < r < a$, as shown in Fig. 3.3. Subsequent unloading produced a counterslip annulus $c < r < a$ and a resultant traction distribution, defined by Eq. (3.26a, b, c), when the tangential force was zero. Reloading from this point will initiate slip at the perimeter of the contact area in the same sense as that produced by the initial loading. This 'reslip' will spread radially inwards over an annulus $d < r < a$ with increasing tangential force and to obtain the distribution of tangential traction during reloading a positive traction of the form

$$q(r) = 2\left(\frac{3\mu F_n}{2\pi a^3}\right)\left[\left(a^2 - r^2\right)^{1/2} - \left(d^2 - r^2\right)^{1/2}\right] \tag{3.31a}$$

$$q(r) = 2\left(\frac{3\mu F_n}{2\pi a^3}\right)\left(a^2 - r^2\right)^{1/2} \tag{3.31b}$$

is superimposed on the traction distribution at the start of reloading given by Eq. (3.26a, b, c). Integrating the resulting traction distribution over the contact area leads to

$$F_t = \mu F_n\left(1 - \frac{b^3}{a^3}\right) - 2\mu F_n\left(1 - \frac{c^3}{a^3}\right) + 2\mu F_n\left(1 - \frac{d^3}{a^3}\right) \tag{3.32}$$

from which

$$d = a\left[1 - \frac{\left(F_t - F_t^{**}\right)}{2\mu F_n}\right] \tag{3.33}$$

where F_t^{**} is the tangential force from which reloading commenced. The relative tangential displacement is given by

$$\begin{aligned}\delta &= \left(\frac{3\mu F_n}{16G^* a}\right)\left(1 - \frac{b^2}{a^2} + 2\frac{c^2}{a^2} - 2\frac{d^2}{a^2}\right)\\ &= \left(\frac{3\mu F_n}{16G^* a}\right)\left[1 - \left(1 - \frac{F_t^*}{\mu F_n}\right)^{2/3} + 2\left(1 - \frac{\left(F_t^* - F_t\right)}{2\mu F_n}\right)^{2/3} - 2\left(1 - \frac{\left(F_t - F_t^{**}\right)}{2\mu F_n}\right)^{2/3}\right]\end{aligned} \tag{3.34}$$

which leads to the tangential stiffness during reloading

$$k_t = \frac{dF_t}{d\delta} = 8G^* a\left[1 - \frac{\left(F_t - F_t^{**}\right)}{2\mu F_n}\right]^{1/3} \tag{3.35}$$

The tangential force is also dependent on the magnitude of the normal force and hence there exists an infinite set of geometrically similar force-displacement curves, each corresponding to a different value of normal force. Figure 3.5a shows two such curves for the case of tangential loading. A theoretical analysis of the behaviour of elastic spheres in contact under varying oblique forces was presented by Mindlin and Deresiewicz (1953). Solutions were given in the form of instantaneous compliances which, due to the dependence on both the current state and the previous loading history, could not be integrated a priori. However, several loading sequences involving variations of both normal and tangential forces were examined from which general procedural rules were identified. Adopting an incremental approach, the procedure is to update the normal force and contact area

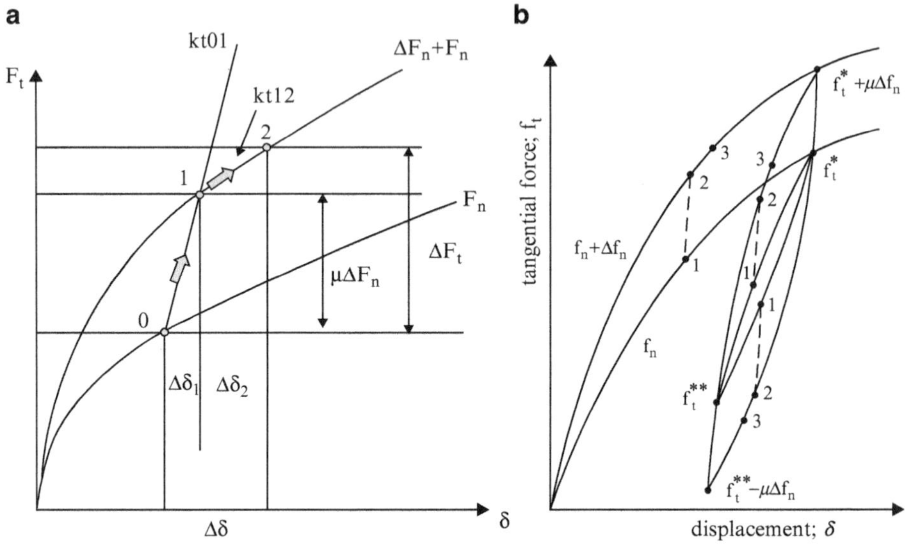

Fig. 3.5 Effect of changing normal force during (**a**) loading (**b**) loading, unloading and reloading

radius, using Eqs. (3.13) and (3.10), followed by calculating ΔF_t using the new values of F_n and a.

From the figure,

$$\Delta\delta = \Delta\delta_1 + \Delta\delta_2 \tag{3.36}$$

where

$$\Delta\delta_1 = \frac{\mu\Delta F_n}{k_{t01}} \quad \text{and} \quad \Delta\delta_2 = \frac{\Delta F_t - \mu\Delta F_n}{k_{t12}} \tag{3.37}$$

with

$$k_{t01} = 8G^*a \ \ and \ \ k_{t12} = 8G^*a\left(1 - \frac{F_t + \mu\Delta F_n}{\mu F_n}\right)^{1/3} \tag{3.38}$$

Hence

$$\Delta\delta = \frac{1}{8G^*a}\left(\mu\Delta F_n + \frac{(\Delta F_t - \mu\Delta F_n)}{\left(1 - \frac{F_t + \mu\Delta F_n}{\mu F_n}\right)^{1/3}}\right) \tag{3.39}$$

By reanalysing all the loading cases considered by Mindlin and Deresiewicz (1953), it was shown by Thornton and Randall (1988) that the tangential incremental displacement may be expressed as

$$\Delta\delta = \frac{1}{8G^* a}\left(\pm\mu\Delta F_n + \frac{\Delta F_t \mp \mu\Delta F_n}{\theta}\right) \tag{3.40}$$

except when

$$\Delta F_n > 0 \text{ and } |\Delta\delta| < \frac{\mu\Delta F_n}{8G^* a} \tag{3.41}$$

Rearrangement of Eq. (3.40) defines the tangential stiffness as

$$k_t = 8G^* a\theta \pm \mu(1-\theta)\frac{\Delta F_n}{\Delta\delta} \tag{3.42}$$

where

$$\theta^3 = 1 - \frac{(F_t + \mu\Delta F_n)}{\mu F_n} \quad \Delta\delta > 0 \text{ (loading)} \tag{3.43a}$$

$$\theta^3 = 1 - \frac{(F_t^* - F_t + 2\mu\Delta F_n)}{2\mu F_n} \quad \Delta\delta < 0 \text{ (unloading)} \tag{3.43b}$$

$$\theta^3 = 1 - \frac{(F_t - F_t^{**} + 2\mu\Delta F_n)}{2\mu F_n} \quad \Delta\delta > 0 \text{ (reloading)} \tag{3.43c}$$

and the negative sign in Eq. (3.42) is only invoked during unloading. The parameters F_t^* and F_t^{**} define the load reversal points, as shown in Fig. 3.5b, and need to be continuously updated

$$F_t^* = F_t^* + \mu\Delta F_n \text{ and } F_t^{**} = F_t^{**} - \mu\Delta F_n \tag{3.44}$$

to allow for the effect of varying normal force.

For a current state given by point 1 in Fig. 3.5b, during loading, unloading or reloading, a tangential incremental displacement equal to the right hand side of Eq. (3.41) will result in a new state given by point 2 on the curve corresponding to the new value of F_n. Larger values of $\Delta\delta$ will result in a state farther along the curve such as point 3. A problem occurs if the conditions given in Eq. (3.41) are true, since point 2 is not reached and the new state does not lie on the curve corresponding to the new value of F_n. However, by adopting an incremental approach, a satisfactory solution to the problem is obtained by setting $\theta = 1$ in Eq. (3.42) until the following condition is satisfied

$$\sum |\Delta F_t| > \mu \sum \Delta F_n \tag{3.45}$$

Note that this problem does not exist for $\Delta F_n < 0$.

3.1.3 Contact Moment

As explained in Sect. 2.1.2, in addition to the normal and tangential force reactions there is also a moment reaction to the twist (rotation about the contact normal vector). An exact solution to this problem was provided by Lubkin (1951) for monotonic loading. The solution is illustrated in Fig. 3.6. Unfortunately the solution is very complex, involving both the definite and indefinite elliptic integrals of the first and second kind; which is not useful to DEM researchers.

By making a small simplifying assumption, Deresiewicz (1954) obtained tangible analytical solutions for both the loading and unloading cases. Figure 3.7a shows the hysteretic moment-rotation curves and Fig. 3.7b shows how the traction distribution changes during unloading. It should be recognized that the moment-rotation problem has all the same complexities as the tangential force-displacement problem defined by Mindlin and Deresiewicz (1953), as described in the previous subsection.

If we define a rotational 'stiffness' k_r then, from Deresiewicz (1954), during loading

$$k_r = \frac{16G^* a^3}{3}\left[2\left(1 - \frac{3}{3\mu F_n a}\frac{M}{}\right)^{-1/2} - 1\right]^{-1} \tag{3.46}$$

and during unloading

$$k_r = \frac{16G^* a^3}{3}\left[2\left(1 - \frac{3}{2}\frac{M^* - M}{2\mu F_n a}\right)^{-1/2} - 1\right]^{-1} \tag{3.47}$$

These equations are for a constant normal force. For a varying normal force there are an infinite number of moment-rotation curves and, for an increase in the normal force, there is a problem if $\Delta M < \mu \Delta F_n a$ which has to be solved like the case of

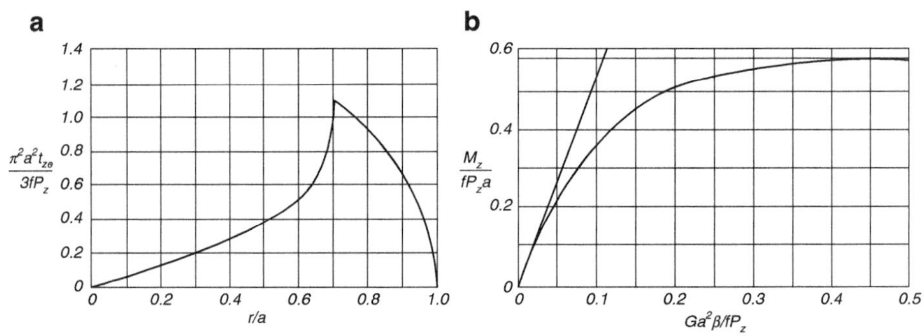

Fig. 3.6 Lubkin's (1951) solution (**a**) torsional traction distribution (**b**) relationship between the twisting moment M and the angle of twist β (Lubkin 1951, Figs. 2 and 3)

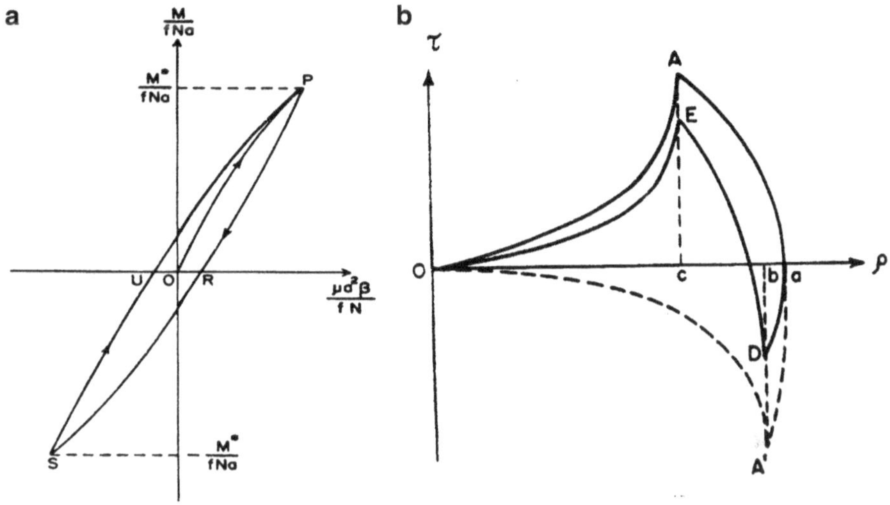

Fig. 3.7 Deresiewicz's (1954) solution (**a**) hysteretic moment-rotation curves (**b**) traction distributions during loading and unloading (Deresiewicz 1954, Figs. 3 and 1)

$\Delta F_t < \mu \Delta F_n$ in the Mindlin and Deresiewicz model, see Sect. 3.1.2. However, to the author's knowledge, the contact moment-rotation model has never been implemented in a DEM code.

3.2 Elastic-Plastic Interactions

With increase in compression between two spheres, ignoring the possibility of fracture or crushing, the material behaviour becomes more complex as it passes through a series of phases, namely elastic, elastoplastic, perfectly plastic and finite plastic deformation (Mesarovic and Fleck 1999). Consequently, for implementation in particle system simulations, some simplifying approximations are inevitable. In this section we ignore the elastoplastic and finite plastic deformation phases and consider that the material stress-strain behaviour is simply linear elastic followed by perfectly plastic, i.e. the stress-strain curve is bilinear.

3.2.1 Normal Interaction

The initial normal interaction is elastic with the normal force, the normal stiffness and the radius of the contact area defined by the theory of Hertz (1881), see Eqs. (3.12), (3.13) and (3.14). Thornton (1997) suggested that the normal interaction becomes plastic when a 'limiting contact pressure' p_y is reached at the centre of

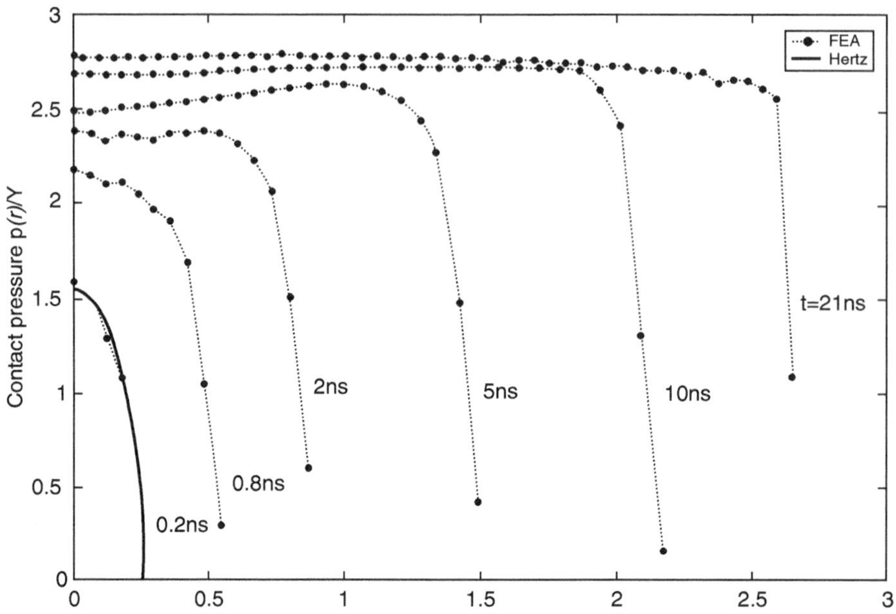

Fig. 3.8 Evolution of the normal pressure distribution during loading

the contact area. This idea originates from Hardy et al. (1971) who reported results of a finite element analysis of a rigid sphere indenting an elastic-perfectly plastic half-space, see Johnson (1985) Fig. 6.11. They showed that the Hertzian pressure distribution is valid until the pressure at the centre of the contact area is equal to 1.6 times the yield stress σ_y of the material, at which point yield occurs below the surface at the centre of the contact area. Further compression results in a spreading of the plastic deformation zone below the surface and a slight modification of the shape of the contact pressure distribution as the maximum contact pressure increases further. When the pressure at the centre of the contact area reaches about 2.4 times the yield stress, the plastic deformation zone in the substrate reaches the contact surface at the perimeter of the contact area. Beyond this point, further compression results in a significant change in the form of the pressure distribution. Over an increasing central portion of the contact area the contact pressure becomes almost constant with only a slight increase in the pressure at the centre of the contact area. This is also illustrated in Fig. 3.8, which shows the evolution of the normal contact pressure distribution for the impact of an elastic sphere against an elastic-perfectly plastic half-space, obtained from finite element analysis by Wu et al. (2003).

Thornton (1997) proposed that the evolution of the contact pressure distribution, described above and illustrated in Fig. 3.8, could be approximated by an 'elastic' stage during which the pressure distribution is Hertzian followed by a 'plastic' stage during which the pressure distribution is described by a truncated Hertzian

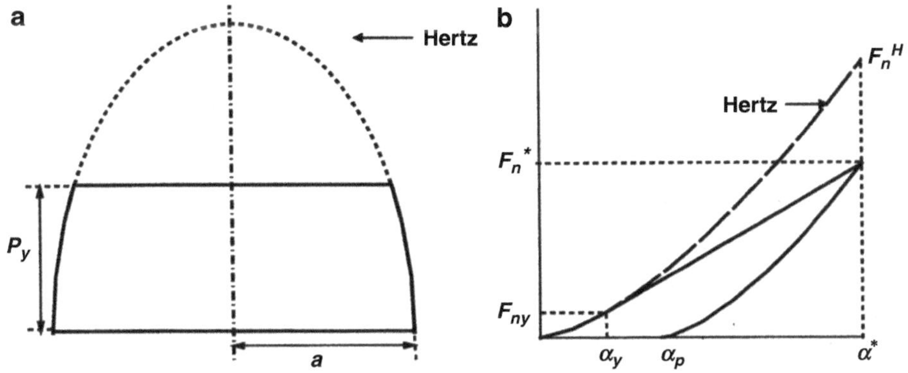

Fig. 3.9 The Thornton (1997) model (**a**) normal pressure distribution (**b**) force-displacement relationship

distribution by defining the limiting contact pressure $p_y \approx 2.5\sigma_y$, as shown in Fig. 3.9a. The consequence of this approximation is that the plastic loading curve is given by a straight line tangent to the Hertzian curve at (F_{ny}, α_y) where F_{ny} and α_y are the normal contact force and relative approach when the pressure at the centre of the contact area first becomes equal to the specified value of p_y, as shown in Fig. 3.9b.

The limiting contact pressure p_y is specified and defined by

$$p_y = \frac{3F_{ny}}{2\pi a_y^2} \tag{3.48}$$

The corresponding relative approach α_y is obtained by substituting Eqs. (3.12) and (3.13) into Eq. (3.48) and rearranging to give

$$\alpha_y = \left(\frac{\pi p_y}{2E^*}\right)^2 R^* \tag{3.49}$$

The corresponding normal force is

$$F_{ny} = \frac{4}{3} E^* R^{*1/2} \alpha_y^{3/2} \tag{3.50}$$

It is assumed that for $\alpha < \alpha_y$ the response is elastic and Hertzian theory is used to calculate the normal force and contact radius. It is also assumed that further compression results in plastic deformation. Therefore, for $\alpha > \alpha_y$, the normal force is given by

$$F_n = F_n^H - 2\pi \int_0^{a_p} \left[p(r) - p_y\right] r dr \tag{3.51}$$

where F_n^H is the equivalent Hertzian force, given by Eq. (3.10), that would result in the same total contact area and a_p is the radius of the contact area over which a uniform pressure p_y is assumed, as indicated in Fig. 3.9a. Integrating Eq. (3.51) we obtain

$$F_n = \pi a_p^2 p_y + F_n^H \left[1 - \left(\frac{a_p}{a}\right)^2\right]^{3/2} \tag{3.52}$$

The limiting contact pressure p_y is defined by Eq. (3.48) or, according to Fig. 3.9a, by

$$p_y = \frac{3F_n^H}{2\pi a^2}\left[1 - \left(\frac{a_p}{a}\right)^2\right]^{1/2} \tag{3.53}$$

The contact radius is obtained from

$$a^3 = \frac{3R^* F_n^H}{4E^*} \tag{3.54}$$

By combining Eqs. (3.48), (3.53) and (3.54) we find that

$$\left[1 - \left(\frac{a_p}{a}\right)^2\right] = \left(\frac{a_y}{a}\right)^2 \quad \text{or} \quad a^2 = a_p^2 + a_y^2 \tag{3.55}$$

Substituting Eqs. (3.54) and (3.55) into Eq. (3.52) we obtain

$$F_n = F_{ny} + \pi p_y \left(a^2 - a_y^2\right) \tag{3.56}$$

Substituting Eq. (3.12), the force-displacement relationship during plastic loading is given as

$$F_n = F_{ny} + \pi R^* p_y (\alpha - \alpha_y) \tag{3.57}$$

which is linear, as shown in Fig. 3.9b.

According to Thornton (1997), the normal force during unloading can be approximated by

$$F_n = \frac{4}{3} E^* R_p^{*1/2} (\alpha - \alpha_p)^{3/2} \tag{3.58}$$

where $R_p^* > R^*$ due to contact flattening and is calculated using

$$R_p^* = \frac{4E^*}{3F_n^*}\left(\frac{2F_n^* + F_{ny}}{2\pi p_y}\right)^{3/2} \tag{3.59}$$

and F_n^* is the maximum normal force from which unloading commenced.

The parameter α_p in Eq. (3.58) is the relative approach at which the normal force becomes zero, i.e. α_p is the plastic (irrecoverable) displacement. It needs to be calculated before Eq. (3.58) can be used. This is achieved by rearranging Eq. (3.58) to give

$$\alpha_p = \alpha^* - \left(\frac{3F_n^*}{4E^* R_p^{*1/2}} \right)^{2/3} \tag{3.60}$$

where α^* is the maximum relative approach at the point of unloading. During unloading the contact radius is calculated from

$$a^3 = \frac{3R_p^* F_n}{4E^*} \tag{3.61}$$

An attraction of this model is that in the context of particle impacts it leads to an analytical solution for the normal coefficient of restitution, see Chap. 5. However, the unloading stage is not strictly correct. Unloading is elastic but not Hertzian as implied by Eq. (3.58). The pressure distribution during unloading for elastic- plastic interactions was analysed by Mesarovic and Johnson (2000). By using rigid punch decomposition, they derived the following equation for the normal pressure distribution during unloading.

$$p = \frac{2p_y}{\pi} sin^{-1} \left(\frac{a^2 - r^2}{a^{*2} - r^2} \right)^{1/2} \tag{3.62}$$

where a^* is the contact radius at the start of unloading. The normal pressure distribution during unloading is illustrated in Fig. 3.10 which shows the analytical solution given by Eq. (3.62) and results of finite element analysis by Wu et al. (2003).

The corresponding normal force relationship, given by Mesarovic and Johnson (2000) is

$$F_n = \frac{2F_n^*}{\pi} \left\{ \sin^{-1}\left(\frac{a}{a^*}\right) - \frac{a}{a^*}\left[1 - \left(\frac{a}{a^*}\right)^2\right]^{1/2} \right\} \tag{3.63}$$

The relative approach during unloading is $\alpha = \alpha^* - \alpha_u$ where α_u is the recovered deformation during unloading. Martin (2003) derived the following equation for α_u.

$$\alpha_u = \frac{2p_y a^*}{E^*} \left[1 - \left(\frac{a}{a^*}\right)^2\right]^{1/2} \tag{3.64}$$

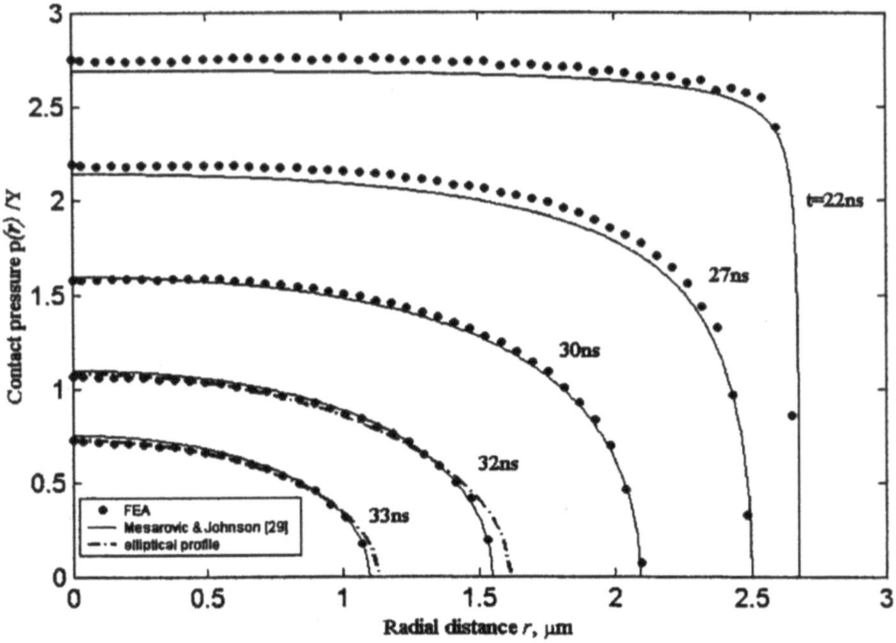

Fig. 3.10 Evolution of the normal pressure distribution during unloading

Substituting Eq. (3.64) into Eq. (3.63) the normal force displacement relationship can be expressed as

$$F_n = \frac{2F_n^*}{\pi}\left[\sin^{-1}\left(1-\lambda^2\right)^{1/2} - \lambda\left(1-\lambda^2\right)^{1/2}\right] \tag{3.65}$$

where

$$\lambda = \frac{E^*(\alpha^* - \alpha)}{2p_y a^*} \tag{3.66}$$

If Thornton's (1997) model is used for the loading stage (any change in R* during plastic loading is ignored) then

$$\alpha^* = \frac{2F_n^* + F_{ny}}{2\pi R^* p_y} \quad \text{and} \quad a^* = \left(\frac{2F_n^* + F_{ny}}{2\pi\, p_y}\right) \tag{3.67}$$

can be used. Figure 3.11 shows a comparison between the unloading prediction using Eq. (3.65) and the assumed Hertzian unloading of Thornton (1997). The

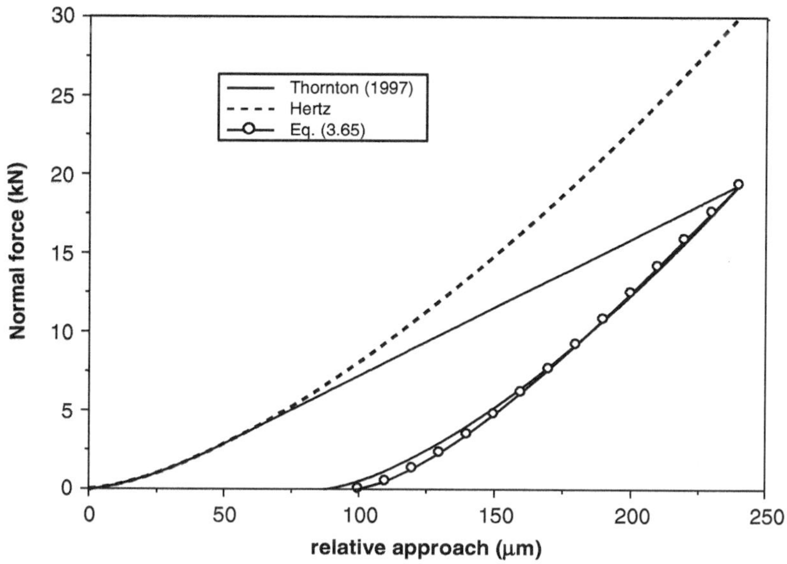

Fig. 3.11 Comparison between the Hertzian and non-Hertzian unloading models

non-Hertzian unloading curve exhibits slightly less elastic recovery but the difference is very small.

3.2.2 *Tangential Interaction and Contact Moment*

It was originally assumed that the tangential interaction is the same as for elastic interactions, i.e. Mindlin and Deresiewicz (1953), but that the tangential stiffness would be higher since, for a given normal force the contact area would be larger. This assumption has been demonstrated to be correct using finite element analysis of the tangential interaction between a rigid sphere and elastic and plastic substrates with the normal force held constant, C-Y Wu (2009, personal communication) unpublished work, as shown in Fig. 3.12. Figure 3.12a clearly shows that the plastic tangential interaction is much stiffer. A further consequence is that sliding occurs at a much smaller value of tangential displacement.

According to Mindlin and Deresiewicz (1953) the tangential force-displacement relationship for a constant normal force may be written as

$$F_t = \mu F_n \left[1 - \left(1 - \frac{\delta}{\delta_\mu} \right)^{3/2} \right] \tag{3.68}$$

where δ_μ is the tangential displacement at which sliding commences and, for the interaction between a rigid sphere and a planar surface, can be expressed as

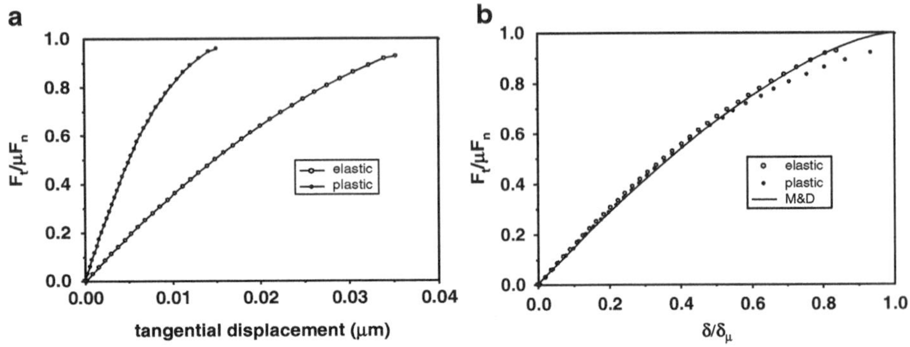

Fig. 3.12 Tangential force-displacement curves (**a**) comparison of elastic and plastic cases (**b**) comparison with the theory of Mindlin and Deresiewicz (1953)

$$\delta_\mu = \frac{3(1+\nu)(2-\nu)}{8aE} \tag{3.69}$$

The data shown in Fig. 3.12a is re-plotted in Fig. 3.12b with Eq. (3.68) superimposed, which confirms the applicability of Mindlin and Deresiewicz (1953) solutions to elastic-plastic tangential interactions.

In a similar way, it is expected that the Deresiewicz (1954) solution for the contact moment will also apply to elastic-plastic interactions, although verification of this is still awaited.

3.3 Adhesive, Elastic Interactions

Particles may adhere together in a number of different ways depending on the type of bond formed. For relatively strong bonds the bond may be solid, cemented or glued by a viscous liquid. Weaker bonds may be provided by pendular liquid bridges, van der Waals forces, electrostatics or electro-magnetic fields. The type of bond dictates the contact interaction law required. However, in this section we only consider autoadhesive particles that result from van der Waals forces.

For particle sizes less than say 50 μm, van der Waals forces become significant and particles tend to stick to each other. Two theoretical models for the autoadhesion of spheres were developed over 40 years ago (a) the JKR model (Johnson et al. 1971) and (b) the DMT model (Derjaguin et al. 1975). The two models were initially thought to be competitive but subsequently shown to be limits to a range of solutions that are governed by the non-dimensional parameter

$$\mu = \left(\frac{R^* \Gamma^2}{E^{*2} z_0^3}\right)^{1/3} \tag{3.70}$$

where z_0 is the equilibrium separation in the Lennard-Jones potential, Γ is the work of adhesion and $\Gamma = \gamma_1 + \gamma_2 - \gamma_{12}$ where γ_1 and γ_2 are the surface energies of the two solids and γ_{12} is the interface energy, i.e. $\Gamma = 2\gamma$ for like spheres. The general consensus is that JKR theory applies for $\mu > 5$ and DMT theory applies for $\mu < 0.1$. For intermediate values of μ a complicated solution has been provided by Maugis (1992).

The maximum tensile force required to break the contact, the so-called 'pull-off' force, is

$$F_{nc} = 1.5\pi\Gamma R^* \tag{3.71}$$

for the JKR model, and

$$F_{nc} = 2\pi\Gamma R^* \tag{3.72}$$

for the DMT model.

It may be noted that the DMT curve is simply the Hertzian curve displaced by a constant amount given by Eq. (3.72). In other words, according to DMT theory, the adhesion does not modify the elastic interaction and the contact breaks when $\alpha = 0$. The JKR theory is more complex.

3.3.1 Normal Interaction

Johnson (1976) provided the following relationship between the normal contact force F_n and the relative approach α., which is shown in Fig. 3.13.

$$\frac{\alpha}{\alpha_f} = \frac{3\left(\frac{F_n}{F_{nc}}\right) + 2 + 2\left(1 + \frac{F_n}{F_{nc}}\right)^{1/2}}{3^{2/3}\left[\frac{F_n}{F_{nc}} + 2 + 2\left(1 + \frac{F_n}{F_{nc}}\right)^{1/2}\right]^{1/3}} \tag{3.73}$$

where

$$\alpha_f = \left(\frac{3F_{nc}^2}{16R^*E^{*2}}\right)^{1/3} \tag{3.74}$$

is the relative approach (negative) at which the contact breaks, point D in Fig. 3.13.

When the two surfaces come into contact the normal force between the two spheres immediately drops to a value of $F_n = -8F_{nc}/9$ (point A in Fig. 3.13) due to van der Waals attractive forces. If the two spheres are compressed until the contact force reaches a maximum (say point B in Fig. 3.13) and then decompressed until $\alpha = 0$ then all the work done during the loading stage has been recovered when point A is reached during unloading. At this point, when $\alpha = 0$, the contact area is

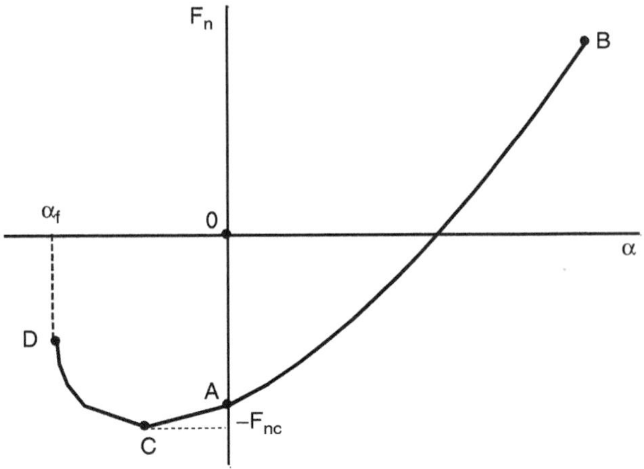

Fig. 3.13 Normal force-displacement curve (JKR theory)

not zero, the spheres remain adhered together, and further work is required to break the contact and create 'new' surface area. If the system is load controlled then the situation becomes unstable at point C when $F_n = -F_{nc}$, $\alpha = -\alpha_f/3^{2/3}$ and the contact breaks. However, particle system simulations are always displacement driven in which case separation occurs at point D when $\alpha = -\alpha_f$ and $F_n = -5F_{nc}/9$.

It was shown by Johnson et al. (1971) that in the presence of adhesion the contact radius a may be defined from

$$a^3 = (3R^*/4E^*)F_n^H \tag{3.75}$$

in which

$$F_n^H = F_n + 2F_{nc} + \left(4F_nF_{nc} + 4F_{nc}^2\right)^{1/2} \tag{3.76}$$

where F_n^H is the apparent Hertzian force required to create the same contact area, which is larger than the actual applied normal force F_n.

The normal contact pressure distribution can be written (Johnson 1985) as

$$p(r) = \left(\frac{2E^*a}{\pi R^*}\right)\left[1 - \left(\frac{r}{a}\right)^2\right]^{1/2} - \left(\frac{2E^*\Gamma}{\pi a}\right)^{1/2}\left[1 - \left(\frac{r}{a}\right)^2\right]^{-1/2} \tag{3.77}$$

or

$$p(r) = \left(\frac{3F_n^H}{2\pi a^2}\right)\left[1 - \left(\frac{r}{a}\right)^2\right]^{1/2} - \left(\frac{F_n^H - F_n}{2\pi a^2}\right)\left[1 - \left(\frac{r}{a}\right)^2\right]^{-1/2} \tag{3.78}$$

where $F_n^H - F_n = \left(8\pi E^*\Gamma a^3\right)^{1/2}$ is the force due to the adhesion traction. Therefore the applied normal contact force is given by

$$F_n = \frac{4E^* a^3}{3R^*} - \left(8\pi E^* \Gamma a^3\right)^{1/2} \tag{3.79}$$

and the corresponding relative approach by

$$\alpha = \frac{a^2}{R^*} - \left(\frac{2\pi \Gamma a}{E^*}\right)^{1/2} \tag{3.80}$$

By differentiating both Eqs. (3.79) and (3.80) with respect to a and then combining to obtain $dF_n/d\alpha$ the normal contact stiffness can be defined by

$$k_n = 2E^* a \left[\frac{3 - 3\left(\frac{a_c}{a}\right)^{3/2}}{3 - \left(\frac{a_c}{a}\right)^{3/2}}\right] \tag{3.81}$$

or

$$k_n = 2E^* a \left(\frac{3\sqrt{F_n^H} - 3\sqrt{F_{nc}}}{3\sqrt{F_n^H} - \sqrt{F_{nc}}}\right) \tag{3.82}$$

Which degenerates to the Hertzian solution ($k_n = 2E^* a$) when there is no adhesion ($F_{nc} = 0$).

3.3.2 Tangential Interaction

Savkoor and Briggs (1977) extended the JKR analysis to account for the effect of oblique loading in the presence of adhesion. It was suggested that the tangential traction distribution over the contact area would be prescribed by the 'no-slip' solution of Mindlin (1949) given by

$$q(r) = \frac{F_t}{2\pi a^2}\left[1 - \left(\frac{r}{a}\right)^2\right]^{-1/2} \tag{3.83}$$

with the relationship between the tangential contact force and the tangential displacement at the contact given by

$$F_t = 8G^* a\delta \tag{3.84}$$

It was also suggested that the application of a tangential force reduces the potential energy by an amount $F_t\delta/2$ leading to the following expression for the radius of the contact area, see Thornton and Yin (1991).

$$a^3 = \frac{3R^*}{4E^*}\left[F_n + 2F_{nc} + \left(4F_nF_{nc} + 4F_{nc}^2 - \frac{F_t^2E^*}{4G^*}\right)^{1/2}\right] \quad (3.85)$$

Equation (3.85) indicates a reduction in the contact radius under increasing tangential force. Savkoor and Briggs (1977) suggested that this corresponds to a 'peeling' mechanism that continues in a stable manner until a critical value F_{tc} is reached when the square root in Eq. (3.85) becomes zero. Hence,

$$F_{tc} = 4\left[\left(F_nF_{nc} + F_{nc}^2\right)G^*/E^*\right]^{1/2} \quad (3.86)$$

It was argued by Savkoor and Briggs (1977) that when $F_t = F_{tc}$ the contact area would collapse to the Hertzian value. However, at $F_t = F_{tc}$ Eq. (3.85) reduces to

$$a^3 = \frac{3R^*}{4E^*}(F_n + 2F_{nc}) \quad (3.87)$$

and it was suggested by Thornton (1991) that, following peeling the micro-slip model of Mindlin and Deresiewicz (1953) could apply, see Sect. 3.1.2, by replacing F_n by $(F_n + 2F_n)$ and using Eq. (3.87) to define the contact radius. It then follows that the sliding condition becomes

$$F_t = \mu(F_n + 2F_{nc}) \quad (3.88)$$

However, although experimental evidence of peeling has been provided for rubber (Savkoor and Briggs 1977) results obtained by Homola et al. (1990) for mica indicated that no peeling occurred. Consequently, we might conclude that the effect of a tangential force in the presence of adhesion is not yet fully understood and may be dependent of the type of material used.

Until this problem is resolved an alternative and simpler model, which might be appropriate for hard materials, is to assume that no peeling process occurs. In this case, the tangential stiffness is

$$k_t = 8G^*a \quad (3.89)$$

where the contact radius a is obtained from Eq. (3.75) and the sliding criterion is

$$F_t = \mu F_n^H \quad (3.90)$$

in which F_n^H is defined by Eq. (3.76).

3.4 Adhesive, Elastic-Plastic Interactions

Two contact mechanics based approaches to the modelling of adhesive, elastic-plastic particle interactions have been reported in the literature by Martin (2003) and by Thornton and Ning (1998).

3.4.1 Normal Interaction

Martin (2003) neglected any adhesion effects during loading and modelled the loading stage as elastic-plastic in a manner very similar to that of Thornton (1997) except that the plastic loading stiffness was higher. During unloading, the repulsive force is given by Eq. (3.65) to which is added the negative adhesion force to obtain

$$F_n = \frac{2F_n^*}{\pi}\left[\sin^{-1}\left(1-\lambda^2\right)^{1/2} - \lambda\left(1-\lambda^2\right)^{1/2}\right] - \left(8\pi\Gamma E^* a^3\right)^{1/2} \tag{3.91}$$

where

$$\lambda = \frac{E^*\left(\alpha^* - \alpha\right)}{2p_y a^*} = \left[1 - \left(\frac{a}{a^*}\right)^2\right]^{1/2} \tag{3.92}$$

Hence

$$a^3 = a^{*3}\left(1-\lambda^2\right)^{3/2} \tag{3.93}$$

and

$$F_n = \frac{2F_n^*}{\pi}\left[\sin^{-1}\left(1-\lambda^2\right)^{1/2} - \lambda\left(1-\lambda^2\right)^{1/2}\right] - \left[8\pi\Gamma E^* a^{*3}\left(1-\lambda^2\right)^{3/2}\right]^{1/2} \tag{3.94}$$

This model was used to simulate powder compacts for which the major concerns are the control of dimensions due to springback during unloading and the strength of the resulting green compact. In this context, the model is satisfactory but, due to the neglect of adhesion during loading, it is not appropriate for general problems. For example, the model cannot be used to simulate agglomeration in which particles stick together when they collide with each other.

A more comprehensive model was presented by Thornton and Ning (1998). In order to model elastic-plastic spheres with adhesion, as in the case of non-adhesive

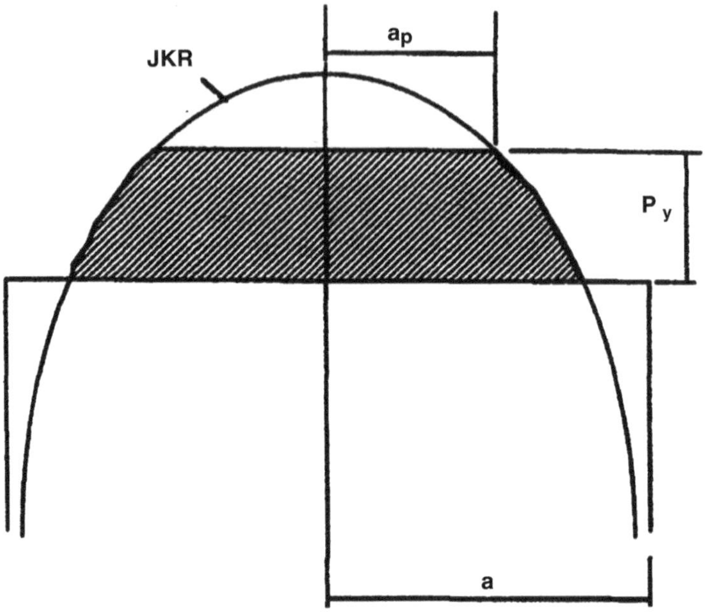

Fig. 3.14 Normal pressure distribution for adhesive, elastic-plastic spheres

elastic-plastic spheres, they again assumed a limiting contact pressure p_y, as shown in Fig. 3.14.

The applied force during plastic deformation, F_{np}, is obtained from

$$F_{np} = F_n - 2\pi \int_0^{a_p} \left[p(r) - p_y \right] r dr \tag{3.95}$$

which leads to

$$\begin{aligned} F_{np} = F_n - \frac{4E^* a^3}{3R^*} \left[1 - \left(1 - \frac{a_p^2}{a^2} \right)^{3/2} \right] \\ + \left(8\pi \Gamma E^* a^3 \right)^{1/2} \left[1 - \left(1 - \frac{a_p^2}{a^2} \right)^{1/2} \right] + \pi a_p^2 p_y \end{aligned} \tag{3.96}$$

with F_n defined by Eq. (3.79). Using Eq. (3.77), the limiting contact pressure is given by

$$p_y = \frac{2E^* a_y}{\pi R^*} - \left(\frac{2\Gamma E^*}{\pi a_y} \right)^{1/2} \tag{3.97}$$

Unfortunately, for the case of adhesive spheres, Eq. (3.55) is not exactly correct but we nevertheless assume it to be a reasonable approximation in order to simplify Eq. (3.96) and obtain

$$F_{np} = \frac{4E^* a_y^3}{3R^*} - a_y (8\pi \Gamma E^{*a})^{1/2} + \pi p_y \left(a^2 - a_y^2 \right) \tag{3.98}$$

Differentiating Eqs. (3.98) and (3.80),

$$\frac{dF_{np}}{da} = 2\pi p_y a - \left(\frac{2\pi \Gamma E^* a_y^2}{a} \right)^{1/2} \tag{3.99}$$

$$\frac{d\alpha}{da} = \frac{2a}{R^*} - \left(\frac{\pi \Gamma}{2E^* a} \right)^{1/2} \tag{3.100}$$

By combining Eqs. (3.99) and (3.100), and substituting Eqs. (3.66) and (3.70), it can be shown that the contact stiffness during plastic loading is given by

$$\frac{dF_{np}}{d\alpha} = \frac{3\pi R^* p_y \sqrt{F_n^H} - 2E^* a_y \sqrt{F_{nc}}}{3\sqrt{F_n^H} - \sqrt{F_{nc}}} \tag{3.101}$$

Note that, for non-adhesive elastic-plastic spheres, $\pi R^* p_y = 2E^* a_y$ in which case Eq. (3.101) would lead to $dF_{np}/d\alpha = \pi R^* p_y$. However, for adhesive elastic-plastic spheres

$$\frac{\pi R^* p_y}{2E^* a_y} = \frac{3\sqrt{F_n^H} - 3\sqrt{F_{nc}}}{3\sqrt{F_n^H} - \sqrt{F_{nc}}} \tag{3.102}$$

During elastic unloading, by analogy with Eq. (3.82), the unloading stiffness is defined by

$$\frac{dF_n}{d\alpha} = 2E^* a \left(\frac{3\sqrt{F_{nr}^H} - 3\sqrt{F_{ncr}}}{3\sqrt{F_{nr}^H} - \sqrt{F_{ncr}}} \right) \tag{3.103}$$

where

$$F_{nr}^H = \frac{4E^* a^3}{3R_p^*} \tag{3.104}$$

and

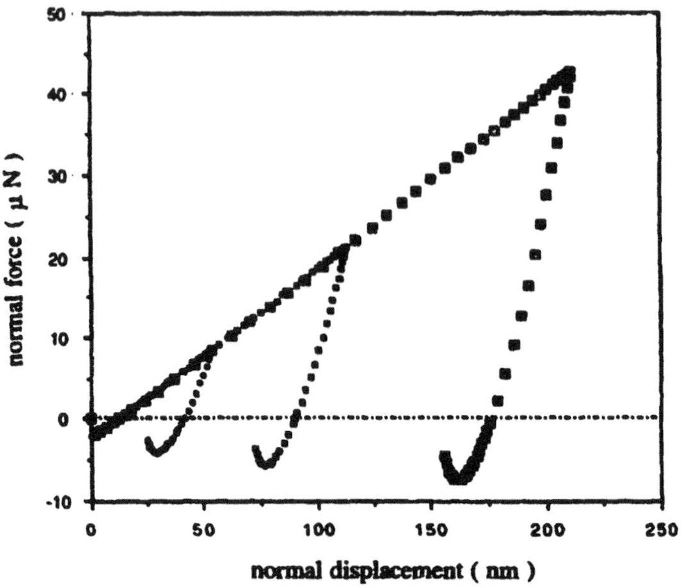

Fig. 3.15 Force-displacement curves

$$F_{ncr} = 1.5\pi \Gamma R_p^* \tag{3.105}$$

with

$$R_p^* = \frac{R^* F_n^{H*}}{F_n^* + \sqrt{4F_{nc}F_n^{H*}}} \tag{3.106}$$

Where F_n^* and F_n^{H*} are the applied load and the equivalent Hertzian force from which unloading commenced. The corresponding force-displacement behaviour for adhesive elastic-plastic interactions is illustrated in Fig. 3.15. The figure demonstrates that the plastic loading curve is linear and that, with increase in initial compression, the unloading stiffness increases and the pull-off force required to overcome adhesion increases due to the decreased radius of curvature caused by plastic flattening during loading.

3.4.2 Tangential Interaction

There is no detailed experimental information about the tangential interaction for adhesive, elastic-plastic spheres. However, for non-adhesive, elastic-plastic spheres, in Sect. 3.2.2 it was demonstrated, Fig. 3.12, that the elastic theory of

Mindlin and Deresiwicz (1953) also applies to elastic-plastic interactions but with a higher stiffness due to the larger contact area. Consequently, in the case of adhesive, elastic-plastic spheres, a strategy that would be expected to produce reasonable results is to apply the theory described in Sect. 3.3.2 for adhesive elastic spheres.

References

Deresiewicz, H.: Contact of elastic spheres under an oscillating torsional couple. Trans. ASME J. Appl. Mech. **76**, 52–56 (1954)

Derjaguin, B.V., Muller, V.M., Toporov, Y.P.: Effect of contact deformations on the adhesion of particles. J. Colloid Interf. Sci. **53**, 314–325 (1975)

Hardy, C., Baronet, C.N., Tordion, G.V.: The elasto-plastic indentation of a half-space by a rigid sphere. Int. J. Numer. Methods Eng. **3**, 451–462 (1971)

Hertz, H.: Über die berűhrung fester elastischer kőrper. J. Reine Angew. Math. **92**, 156–171 (1881)

Homola, A.M., Israelachvili, J.N., McGuiggan, P.M., Hellgeth, J.W.: Fundamental studies in tribology. Wear **136**, 65–84 (1990)

Johnson, K.L.: Adhesion at the contact of solids. In: Koiter, W.T. (ed.) Theoretical and Applied Mechanics. Proc. 4th IUTAM Congress, pp. 133–143. North-Holland, Amsterdam (1976)

Johnson, K.L.: Contact Mechanics. Cambridge University Press, Cambridge (1985)

Johnson, K.L., Kendall, K., Roberts, A.D.: Surface energy and contact of elastic solids. Proc. R. Soc. Lond. **A324**, 301–313 (1971)

Lubkin, J.L.: The torsion of elastic spheres in contact. Trans. ASME J. Appl. Mech. **73**, 183–187 (1951)

Martin, C.L.: Unloading of powder compacts and their resulting tensile strength. Acta Mater. **51**, 4589–4602 (2003)

Maugis, D.: Adhesion of spheres: the JKR-DMT transition using a Dugdale model. J. Colloid Interf. Sci. **150**, 243–269 (1992)

Mesarovic, S.D., Fleck, N.A.: Spherical indentation of elastic-plastic solids. Proc. R. Soc. Lond. **A455**, 2707–2728 (1999)

Mesarovic, S.D., Johnson, K.L.: Adhesive contact of elastic-plastic spheres. J. Mech. Phys. Solids **48**, 2009–2033 (2000)

Mindlin, R.D.: Compliance of elastic bodies in contact. Trans. ASME J. Appl. Mech. **16**, 259–268 (1949)

Mindlin, R.D., Deresiewicz, H.: Elastic spheres in contact under varying oblique forces. Trans. ASME J. Appl. Mech. **20**, 327–344 (1953)

Savkoor, A.R., Briggs, G.A.D.: The effect of tangential force on the contact of elastic solids in adhesion. Proc. R. Soc. Lond. **A359**, 103–114 (1977)

Thornton, C.: Interparticle sliding in the presence of adhesion. J. Phys. D Appl. Phys. **24**, 1942–1946 (1991)

Thornton, C.: Coefficient of restitution for collinear collisions of elastic-perfectly plastic spheres. Trans. ASME J. Appl. Mech. **64**, 383–386 (1997)

Thornton, C., Ning, Z.: A theoretical model for the stick/bounce behaviour of adhesive, elastic-plastic spheres. Powder Technol. **99**, 154–162 (1998)

Thornton, C., Randall, C.W.: Application of theoretical contact mechanics to solid particle system simulation. In: Satake, M., Jenkins, J.T. (eds.) Micromechanics of Granular Materials, pp. 133–142. Elsevier, Amsterdam (1988)

Thornton, C., Yin, K.K.: Impact of elastic spheres with and without adhesion. Powder Technol. **65**, 153–166 (1991)

Wu, C.Y., Li, L.Y., Thornton, C.: Rebound behaviour of spheres for plastic impacts. Int. J. Impact Eng. **28**, 929–946 (2003)

Chapter 4
Other Contact Force Models

Abstract Implementation of the contact force models described in the previous chapter into particle system simulation codes involve complex algorithms and require load reversal points to be stored in memory. Consequently, many researchers elect to use simpler contact force models in order to reduce computer time requirements. A selection of the most common simpler contact force models are described in this chapter, including linear springs and non-linear springs, with and without dashpots, partially latching springs and adhesive piecewise linear models.

Implementation of the contact force models described in the previous chapter into particle system simulation codes involve complex algorithms and require load reversal points to be stored in memory. Consequently, many researchers elect to use simpler contact force models in order to reduce computer time requirements. A selection of the most common simpler contact force models are described in this chapter.

4.1 Linear Spring Models

The simplest elastic contact force model is to assume that, during contact, two interacting bodies are connected, both normally and tangentially, by linear springs. Hence, the normal and tangential contact forces can be calculated from the following equations

$$F_n = k_n \alpha \tag{4.1}$$

$$F_t^{new} = F_t^{old} + k_t \Delta\delta \text{ except if } F_t^{new} \geq \mu F_n \text{ then } F_t^{new} = \mu F_n \tag{4.2}$$

where k_n and k_t are the normal and tangential spring stiffnesses, α is the relative normal approach and Δδ is the relative tangential surface displacement increment at the contact. Note that the tangential contact force is calculated incrementally in Eq. (4.2) In all contact force models, the tangential interaction has to be calculated incrementally because sliding may have previously occurred and δ is no longer simply the extension of the spring. However, the exception to this rule is the linear spring model. Tsuji et al. (1992) calculated the tangential force using $F_t = k_t\delta$ and

C. Thornton, *Granular Dynamics, Contact Mechanics and Particle System Simulations*, Particle Technology Series 24, DOI 10.1007/978-3-319-18711-2_4

suggested that, when sliding occurred, not only should the tangential force be reset to the Coulomb limit $F_t = \mu F_n$ but also the tangential displacement should be reset to $\delta = \mu F_n / k_t$. The same suggestion was made by Brendel and Dippel (1998). Although the method is correct for a linear spring model, Eq. (4.2) is simpler. For non-linear spring models an incremental approach is essential, see discussion provided by Thornton et al. (2011).

In the context of particle impact, the linear spring model predicts the contact duration t_c to be

$$t_c = \pi \left(\frac{m^*}{k_n} \right)^{1/2} \tag{4.3}$$

with

$$\frac{1}{m^*} = \frac{1}{m_1} + \frac{1}{m_2} \tag{4.4}$$

where m_1 and m_2 are the masses of the two interaction spheres. Unrealistically, the contact duration is independent of the impact velocity. However, it was shown by Thornton et al. (2011) that, if the normal spring stiffness is calibrated by equating Eq. (4.3) to the corresponding contact duration for a Hertzian spring, see Eq. (5.5), very accurate predictions of the rebound characteristics for oblique impacts can be obtained provided that a realistic value is used for the ratio of the normal and tangential contact stiffnesses. As will be explained in the next section, the range of realistic values is $1 \geq k_t/k_n \geq 2/3$.

In the context of quasi-static deformation of compact particle systems with enduring contacts it is not possible to reliably calibrate the spring stiffness. This may be possible for a given stress level, i.e. shearing with the mean stress held constant, but if a range of stress levels are examined then the resultant stress strain curves can become unrealistic.

The most common contact force model, used for both normal and tangential interactions, is the linear spring-dashpot model introduced by Walton (1983). This model is widely used to artificially dissipate energy through viscous means without implying that the particles are actually viscoelastic. In the linear spring-dashpot model the normal force is calculated using

$$F_n = k_n \alpha + 2\gamma \sqrt{m^* k_n} v_n \tag{4.5}$$

where v_n is the relative normal velocity. The tangential force is obtained from

$$F_{te}^{i} = F_{te}^{i-1} + k_t \Delta\delta \tag{4.6}$$

and

$$F_t = F_{te}^{i} + 2\gamma \sqrt{m^* k_t} v_t \quad \text{except if } F_t \geq \mu F_n \quad \text{then } F_t = \mu F_n \tag{4.7}$$

where, for a given timestep i, F_{te}^{i} is the elastic component of the tangential force and v_t is the relative tangential surface velocity. In Eqs. (4.5) and (4.7) the value of the damping coefficient γ is dependent on the desired value of the normal coefficient of restitution e_n which is defined by

$$e_n = \exp\left(-\sqrt{\frac{k_n}{2m^*}}\gamma\, t_c\right) \tag{4.8}$$

where t_c is the contact time during an impact. For the linear spring dashpot model a widely used equation is

$$\gamma = \frac{-\ln e_n}{\sqrt{\pi^2 + \ln^2 e_n}} \tag{4.9}$$

However, the equation is incorrect since it is based on the assumption that the contact time t_c ends when the displacement α first returns to zero. At this point the normal force is tensile which is not acceptable in simulations of non-adhesive particles. For non-adhesive spheres, the appropriate value of t_c to be substituted into Eq. (4.8) is the time when the normal contact force first returns to zero. The correct solution for the linear spring model was provided by Schwager and Pöschel (2007). Their solutions, given by Eq. (23) in their paper, can be rewritten as

$$\ln e_n = -\frac{\gamma}{\sqrt{1-\gamma^2}}\left[\pi - \arctan\left(\frac{2\gamma\sqrt{1-\gamma^2}}{2\gamma^2-1}\right)\right] \quad \text{for} \quad \gamma \le \frac{1}{\sqrt{2}} \tag{4.10a}$$

$$\ln e_n = -\frac{\gamma}{\sqrt{1-\gamma^2}}\arctan\left(\frac{2\gamma\sqrt{1-\gamma^2}}{2\gamma^2-1}\right) \quad \text{for} \quad 1 \ge \gamma \ge \frac{1}{\sqrt{2}} \tag{4.10b}$$

$$\ln e_n = -\frac{\gamma}{\sqrt{1-\gamma^2}}\ln\left(\frac{\gamma+\sqrt{\gamma^2-1}}{\gamma-\sqrt{\gamma^2-1}}\right) \quad \text{for} \quad \gamma \ge 1 \tag{4.10c}$$

It is worth noting that for $\gamma < 0.707$ the coefficient of restitution $e_n > 0.21$ and hence, in particle system simulations, only Eq. (4.10a) is relevant. Using an elaborate curve fitting technique for Eq. (4.10a, b and c), an expression for γ as a function of e_n was provided by Thornton et al. (2013).

Many researchers mistakenly attribute the linear spring-dashpot model to Cundall and Strack (1979). Although dashpots were used by Cundall and Strack (1979) they did not contribute to the contact forces. The contact forces were simply the forces in the springs. However, the dashpot forces were added to the spring forces to provide the contribution to the particle out-of-balance force to be used when calculating the particle accelerations. The original purpose of the dashpots was to suppress 'rattling' at contacts during 2D quasi-static simulations. However, the use of dashpots in the sense as used by Cundall and Strack has a physical justification since they are used to dissipate a small amount of energy due to elastic wave propagation through a solid particle, as explained in Sect. 2.1. When two adhesive spheres collide at small enough relative impact velocities that they stick together then, without these dashpots, the normal and tangential contact forces oscillate ad infinitum and never reach equilibrium.

4.2 Non-linear Spring Models

In order to avoid the complexity of the hysteretic tangential force model of Mindlin and Deresiewicz (1953) a number of researchers choose to use the 'no-slip' model of Mindlin (1949) which, when combined with Hertzian theory for the normal force provides a non-linear spring model; to which non-linear dashpots may be added in order to dissipate energy. Considering the combined non-linear spring-dashpot model the normal and tangential contact forces at the ith timestep are obtained from

$$F_n = \frac{4}{3}E^* \sqrt{R^* \alpha^3} + 2\gamma \sqrt{m^* k_n} v_n \tag{4.11}$$

$$F^i_{te} = F^{i-1}_{te} + k^i_t \Delta\delta \quad \text{for} \quad \Delta F_n \geq 0 \tag{4.12}$$

$$F^i_{te} = F^{i-1}_{te} \left(\frac{k^i_t}{k^{i-1}_t} \right) + k^i_t \Delta\delta \quad \text{for} \quad \Delta F_n < 0 \tag{4.13}$$

$$F_t = F^i_{te} + 2\gamma \sqrt{m^* k_t} v_t \quad \text{except if} \quad F_t \geq \mu F_n \quad \text{then} \quad F_t = \mu F_n \tag{4.14}$$

in which the variable stiffnesses are

$$k_n = 2E^* \sqrt{R^* \alpha} \quad \text{and} \quad k_t = 8G^* \sqrt{R^* \alpha} \tag{4.15}$$

and the damping coefficient γ can be obtained as a function of the desired value of restitution coefficient using Eq. (B6) provided by Thornton et al. (2013)

As indicated by Eqs. (4.12) and (4.13), the elastic component of the tangential force F^i_{te} depends on whether the normal force is increasing or decreasing. If the normal force decreases the contact area decreases. Associated with the reduced contact area is a reduced stiffness and for that stiffness the current magnitude of the relative surface displacement is insufficient to generate the magnitude of the elastic component of the previous tangential force F^{i-1}_{te} which needs to be rescaled according to Eq. (4.13). Elata and Berryman (1996) demonstrated that if the old tangential force is not rescaled then spurious energy can be created.

Using Eqs. (3.6) and (3.21) and the relationship $E = 2(1 + \nu)G$, the contact stiffness ratio κ can be written as

$$\kappa = \frac{k_t}{k_n} = 2\left[\frac{G_2(1-\nu_1)+G_1(1-\nu_2)}{G_2(2-\nu_1)+G_1(2-\nu_2)}\right] \tag{4.16}$$

Note that if $\nu_1 = \nu_2 = \nu$, Eq. (4.16) reduces to

$$\kappa = \frac{2(1-\nu)}{(2-\nu)} \tag{4.17}$$

even if $G_1 \neq G_2$. Also, since $0 \leq \nu \leq 1/2$ the range of realistic values of κ for elastic interactions is $1 \geq \kappa \geq 2/3$; and this constraint also applies to the linear spring model discussed in Sect. 4.1.

The 'no-slip' model of Mindlin (1949) can also be combined with the elastic-plastic normal contact force model of Thornton (1997) to provide energy dissipation without the need for dashpots. In addition, the calculation of the contact moment is simplified since the rotational 'stiffness' k_r given by Eq. (3.46) reduces to

$$k_r = \frac{16G^* a^3}{3} \tag{4.18}$$

4.3 Partially Latching Spring Models

In order to mimic plastic deformation, as observed in experiments, Walton and Braun (1986) and Walton (1993) introduced a so-called 'partially latching spring' model that used different normal spring stiffnesses for loading and unloading. In this model the normal force is written as

$$F_n = k_1 \alpha \tag{4.19}$$

during loading, and

$$F_n = k_2(\alpha - \alpha_0) \tag{4.20}$$

during unloading, where α_0 is the relative approach when the unloading curve reaches zero force and

$$\frac{k_1}{k_2} = e_n^2 \tag{4.21}$$

$$\alpha_0 = \alpha_{max}(1 - k_1/k_2) = \alpha_{max}(1 - e_n^2) \tag{4.22}$$

where e_n is the normal coefficient of restitution and α_{max} is the maximum relative approach from which unloading occurred. A problem with the partially latching spring model, defined by Eqs. (4.19), (4.20), (4.21) and (4.22), is that Eq. (4.21) is

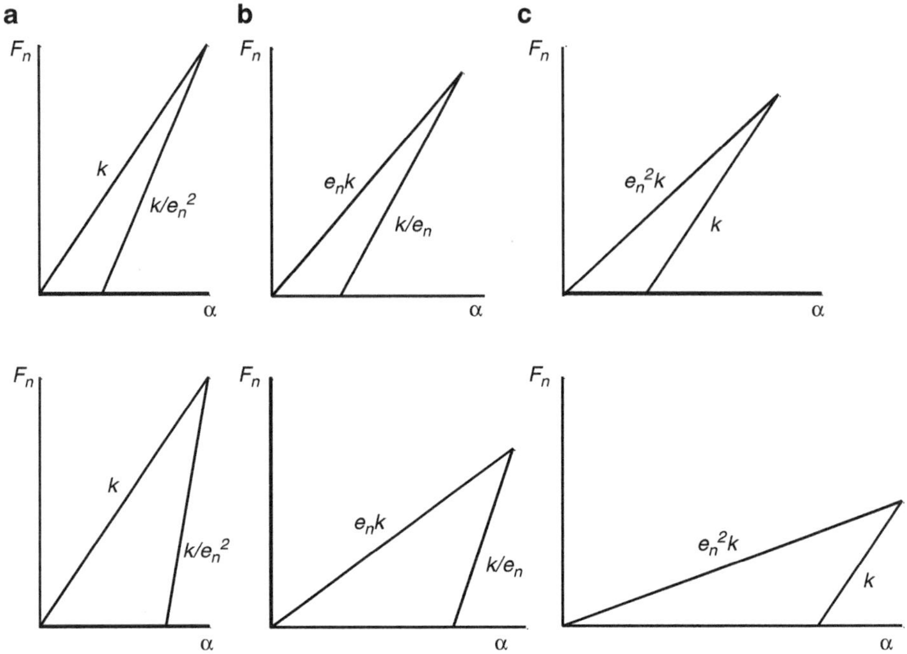

Fig. 4.1 Three possible partially latching spring models for $e_n = 0.8$ (*upper figures*) and $e_n = 0.5$ (*lower figures*)

ambiguous. There are at least three interpretations which are illustrated in Fig. 4.1 and defined below.

Model A is illustrated by Fig. 4.1a

$$F_n = k_1\alpha = k\alpha \text{ during loading} \tag{4.23}$$

$$F_n = k_2(\alpha - \alpha_0) = \frac{k}{e_n^2}(\alpha - \alpha_0) \text{ during unloading} \tag{4.24}$$

$$F_t^i = F_t^{i-1} + \kappa\frac{k}{e_n^2}\Delta\delta \quad \text{except if} \quad F_t \geq \mu F_n \quad \text{then} \quad F_t = \mu F_n \tag{4.25}$$

Model B is illustrated in Fig. 4.1b

$$F_n = k_1\alpha = e_nk\alpha \text{ during loading} \tag{4.26}$$

$$F_n = k_2(\alpha - \alpha_0) = \frac{k}{e_n}(\alpha - \alpha_0) \text{ during unloading} \tag{4.27}$$

$$F_t^i = F_t^{i-1} + \kappa\frac{k}{e_n}\Delta\delta \quad \text{except if} \quad F_t \geq \mu F_n \quad \text{then} \quad F_t = \mu F_n \tag{4.28}$$

Model C is illustrated in Fig. 4.1c

$$F_n = k_1\alpha = e_n^2 k\alpha \text{ during loading} \tag{4.29}$$

$$F_n = k_2(\alpha - \alpha_0) = k(\alpha - \alpha_0) \text{ during unloading} \tag{4.30}$$

$$F_t^i = F_t^{i-1} + \kappa k \Delta\delta \text{ except if } F_t \geq \mu F_n \text{ then } F_t = \mu F_n \tag{4.31}$$

In all three models, k is the elastic stiffness that would result if $e_n = 1.0$.

It was demonstrated by Thornton et al. (2013) that, for oblique impacts, all three models predict exactly the same rebound kinematics when the tangential spring stiffness is related to the normal unloading stiffness, as above. The differences are in the magnitude of the contact forces and the contact duration. Compared to the model of Thornton (1997), see Sect. 3.2.1, Model A significantly over-predicts the normal force and under-predicts the contact duration, whereas Model C significantly under-predicts the force and over-predicts the duration of contact. It was found that the force evolution for Model B was quite similar to that of Thornton's (1997) model due to the fact that the model captures two significant aspects of the model of Thornton (1997) in that for a decrease in e_n the value of the limiting contact pressure decreases and therefore the plastic loading stiffness decreases as indicated by Eq. (3.57) and the unloading stiffness increases since a decrease in p_y leads to an increase in R_p^*, as indicated by Eq. (3.59).

As indicated by Eq. (4.21), all three above models predict a normal coefficient of restitution that is independent of impact velocity. Walton and Braun (1986) suggested a variable latching spring model by making the normal unloading stiffness, as given in Eq. (4.20), a function of the maximum normal force from which unloading commenced, i.e.

$$k_2 = k_1 + sF_{n(max)} \tag{4.32}$$

where s is an empirical scaling coefficient. Using Eq. (4.32) the normal coefficient of restitution is given by

$$e_n = \left(\frac{\sqrt{k_1/m^*}}{\sqrt{k_1/m^*} + sV_{ni}} \right)^{1/2} \tag{4.33}$$

which exhibits a dependency on the normal impact velocity V_{ni}

4.4 Adhesive Piecewise Linear Models

Luding (2005) purported to extend his previous partly latching spring model (Luding 1998) to account for adhesive/cohesive contacts. The corresponding force-displacement law is illustrated in Fig. 4.2a. A more elaborate model was proposed by Luding (2008), which is illustrated in Fig. 4.2b. It is clear that both models are physically unrealistic since they imply that all the deformation produced

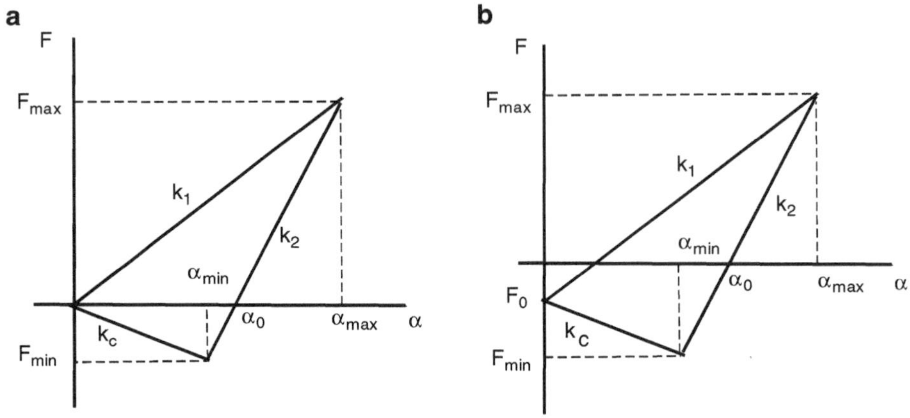

Fig. 4.2 Force-displacement laws (**a**) Luding (2005) (**b**) Luding (2008)

by plastic loading, α_{max}, can be recovered and that contact ceases when $\alpha = 0$. The consequences of this will be illustrated later.

In the Luding (2005) model shown in Fig. 4.2a the force-displacement equations are

$$F = k_1 \alpha \text{ for initial loading} \tag{4.34a}$$

$$F = k_2(\alpha - \alpha_0) \text{ for unloading/reloading} \tag{4.34b}$$

$$F = -k_c \alpha \text{ for further unloading} \tag{4.34c}$$

with

$$\alpha_0 = \left(1 - \frac{k_1}{k_2}\right) \alpha_{max} \tag{4.35}$$

and

$$\alpha_{min} = \left(\frac{k_2 - k_1}{k_2 + k_c}\right) \alpha_{max} \tag{4.36}$$

If we consider a normal impact between a sphere and a planar target wall the initial kinetic energy is given by

$$\frac{1}{2} m V_i^2 = \frac{1}{2} k_1 \alpha_{max}^2 \tag{4.37}$$

and the rebound kinetic energy is given by

$$\frac{1}{2}mV_r^2 = \frac{1}{2}\frac{k_1^2}{k_2}\alpha_{max}^2 - \frac{1}{2}k_c\alpha_0\alpha_{min} \tag{4.38}$$

where the first term on the RHS is the kinetic energy recovered when $\alpha = \alpha_0$ and the second term is the work done when the contact force becomes tensile. Substituting Eqs. (4.35) and (4.36) we obtain

$$\frac{1}{2}mV_r^2 = \frac{1}{2}\frac{k_1^2}{k_2}\alpha_{max}^2 - \frac{1}{2k_2}\frac{k_c}{(k_2+k_c)}\frac{(k_2-k_1)^2}{}\alpha_{max}^2 \tag{4.39}$$

Dividing Eq. (4.39) by Eq. (4.37) leads to

$$e^2 = \frac{k_1}{k_2} - \frac{k_c}{k_1k_2}\frac{(k_2-k_1)^2}{(k_2+k_c)} \tag{4.40}$$

where e is the coefficient of restitution. If $k_c = 0$ and $k_1 = 200$ N/m, $k_2 = 500$ N/m then $e = 0.6325$. For the same values of k_1 and k_2 but with $k_c = 100$ N/m the value of $e = 0.5$. In both cases the coefficient of restitution is independent of impact velocity, even when $V_i \to 0$.

Many experiments have demonstrated that the coefficient of restitution is a function of the impact velocity. In order to obtain a velocity dependent coefficient of restitution it was suggested by Walton and Braun (1986) that the unloading stiffness should be a function of the maximum force, see Eq. (4.32). Since the loading stiffness is linear this is equivalent to suggesting that k_2 is a function of α_{max}. Luding (2008) suggested the following expression.

$$k_2 = k_1 + (\hat{k}_2 - k_1)\frac{\alpha_{max}}{\alpha_{max}^*} \tag{4.41}$$

where

$$\alpha_{max}^* = \frac{\hat{k}_2}{\hat{k}_2 - k_1}\phi_f\frac{2R_1R_2}{R_1+R_2} \tag{4.42}$$

in which R_i are the radii of the two spheres in contact and ϕ_f was described as the "dimensionless plasticity depth". For a sphere impacting a target wall the equation reduces to

$$\alpha_{max}^* = \frac{\hat{k}_2}{\hat{k}_2 - k_1}\phi_f d_p \tag{4.43}$$

Luding (2008) suggested that $\hat{k}_2$ is a limiting unloading stiffness that applies for all $\alpha_{max} \geq \alpha_{max}^*$ and suggested a value of $\phi_f = 0.05$. This is not necessary and, therefore, we will consider $\hat{k}_2$, which is arbitrarily selected, to be the unloading stiffness when $\alpha_{\max} = \hat{\alpha}_{\max}$ and choose $\hat{k}_2 = 500$ N/m and $\phi_f = 0.01$. Consequently, we use

$$\hat{\alpha}_{max} = \frac{\hat{k}_2}{\hat{k}_2 - k_1} \phi_f d_p \tag{4.44}$$

and

$$k_2 = k_1 + (\hat{k}_2 - k_1) \frac{\alpha_{max}}{\hat{\alpha}_{max}} \tag{4.45}$$

Rather than consider the Luding (2008) model as a separate issue we can adapt the Luding (2005) model in order to examine the effects of the various details. In this sense, we examine three different cases in which the unloading stiffness is defined by Eq. (4.45). In all cases we take $\hat{k}_2 = 500$ N/m.

First, we consider the case of no adhesion, i.e. the contact terminates when $\alpha = \alpha_0$ during unloading. Secondly, we consider the case with 'adhesion' in the sense that there is a tensile force permitted, as illustrated in Fig. 4.2a. In the third case, we add an initial negative force F_0, as illustrated in Fig. 4.2b. The results obtained are shown in Fig. 4.3.

The figure shows that for no adhesion the coefficient of restitution reduces at a decreasing rate as the impact velocity increases. For the second case, since a tensile force is admitted the results show that as $V_i \to 0$, $e_n \to 1$ which is counterintuitive. Furthermore, in this case, the coefficient of restitution reduces until, for $V_i > 1.8$ m/s

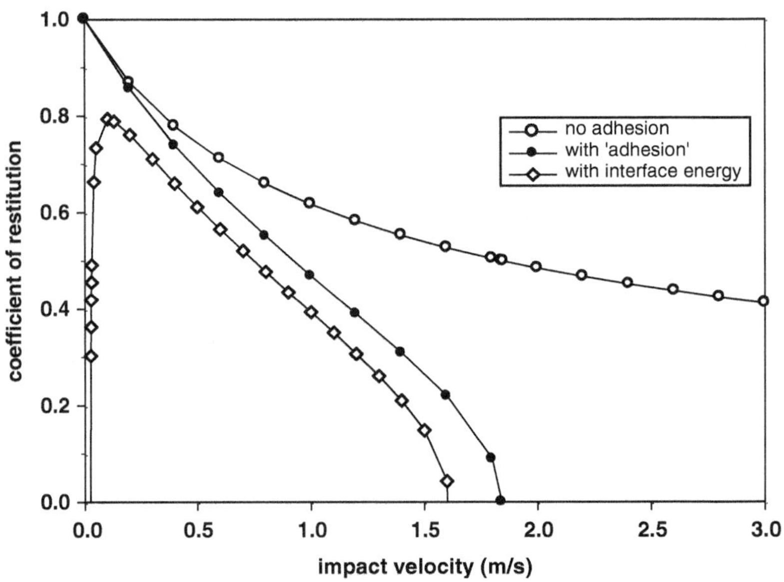

Fig. 4.3 Coefficients of restitution for k_2 as a function of α

there appears to be sticking. However, at these higher velocities sticking does not occur. It is simply the fact that, due to the increasing work done by the tensile force stretching the surfaces back to $\alpha = 0$, there is no energy left for the sphere to rebound from the target wall. It simply sits on the wall but is not sticking to the wall. In other words, with this model there is no adhesion. It is concluded that the results shown in Fig. 4.3 clearly demonstrate that the Luding (2005, 2008) models are physically unrealistic.

For the non-adhesive case a further problem exists, which also applies to the partially latching spring model of Walton and Braun (1986). If tensile forces are not permitted then Eq. (4.40) reduces to

$$e = \sqrt{\frac{k_1}{k_2}} \tag{4.46}$$

If k_2 is a function of α, as given by Eq. (4.45)

$$e^2 = \frac{k_1}{k_1 + \left(\hat{k}_2 - k_1\right)\frac{\alpha_{max}}{\hat{\alpha}_{max}}} = \frac{1}{1 + A\alpha_{max}} \tag{4.47}$$

However,

$$\alpha_{max} = \sqrt{\frac{m}{k_1}} V_i \tag{4.48}$$

and therefore, for high velocities, $e \propto V_i^{-1/2}$ contrary to the well-established fact that for perfect plastic interactions (Johnson 1985; Thornton 1997) $e \propto V_i^{-1/4}$. Consequently, it is clear that, in order to obtain $e \propto V_i^{-1/4}$, k_2 should be a function of $\sqrt{\alpha_{max}}$, i.e.

$$k_2 = k_1 + \left(\hat{k}_2 - k_1\right)\sqrt{\frac{\alpha_{max}}{\hat{\alpha}_{max}}} \tag{4.49}$$

The corresponding results obtained from simulations using Eq. (4.49) are shown in Fig. 4.4.

Pasha et al. (2014) recognised the physically unrealistic nature of Luding's (2008) model and suggested the force-displacement model shown in Fig. 4.5, which is clearly more realistic than that of Luding (2008). However, the unloading stiffness is a function of α and hence the velocity dependence is incorrect.

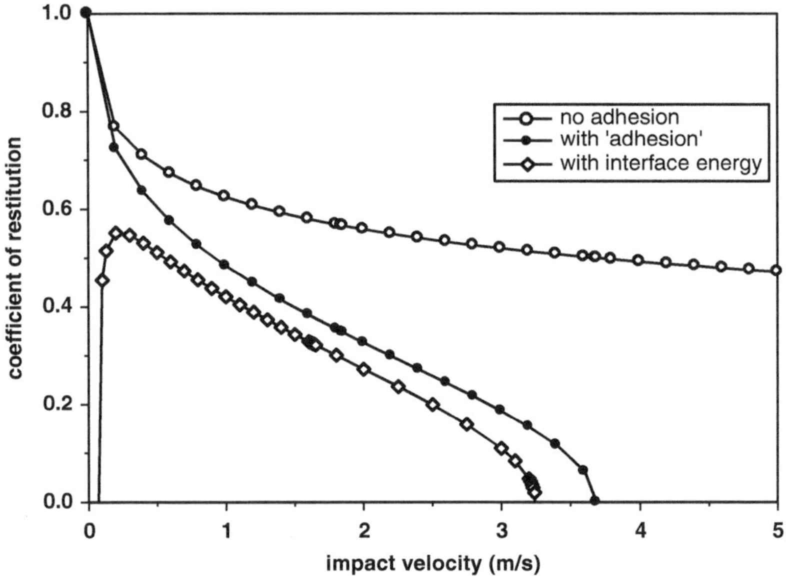

Fig. 4.4 Coefficients of restitution for k_2 as a function of $\sqrt{\alpha}$

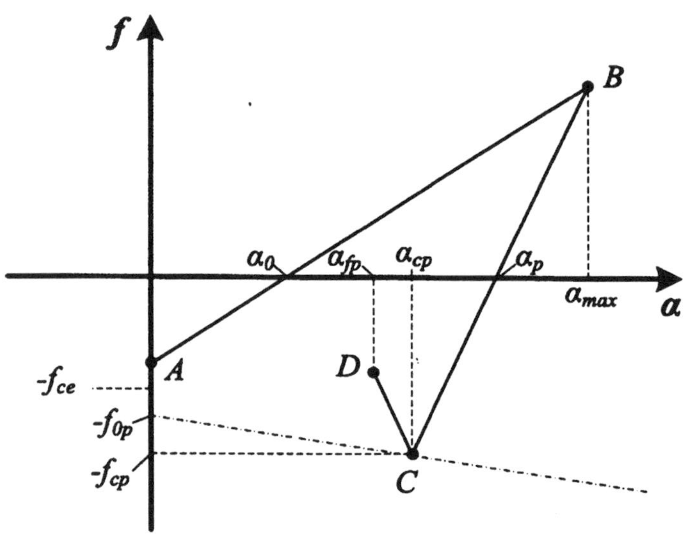

Fig. 4.5 Force-displacement law (Pasha et al. 2014, Fig. 10)

References

Brendel, L., Dippel, S.: Lasting contacts in molecular dynamics simulations. In: Herrmann, H.J., Hovi, J.-P., Luding, S. (eds.) Physics of dry granular media, pp. 313–318. Kluwer Academic Publishers, Dordrecht (1998)

Cundall, P.A., Strack, O.D.L.: A discrete numerical model for granular assemblies. Géotechnique **29**, 47–65 (1979)

Elata, D., Berryman, J.G.: Contact force displacement laws and the mechanical behaviour of random packs of identical spheres. Mech. Mater. **24**, 229–240 (1996)

Johnson, K.L.: Contact Mechanics. Cambridge University Press, Cambridge (1985)

Luding, S.: Collisions and contacts between two particles. In: Herrmann, H.J., Hovi, J.-P., Luding, S. (eds.) Physics of Dry Granular Media, pp. 285–304. Kluwer Academic Publs, Dordrecht (1998)

Luding, S.: Shear flow modelling of cohesive and frictional fine powder. Powder Technol. **158**, 45–50 (2005)

Luding, S.: Cohesive, frictional powders: contact models for tension. Granul. Matter **10**, 235–246 (2008)

Mindlin, R.D.: Compliance of elastic bodies in contact. Trans. ASME J. Appl. Mech. **16**, 259–268 (1949)

Mindlin, R.D., Deresiewicz, H.: Elastic spheres in contact under varying oblique forces. Trans. ASME J. Appl. Mech. **20**, 327–344 (1953)

Pasha, M., Dogbe, S., Hare, C., Hassanpour, A., Ghadiri, M.: A linear model of elasto-plastic and adhesive contact deformation. Granul. Matter **16**, 151–162 (2014)

Schwager, T., Pöschel, T.: Coefficient of restitution and linear-dashpot revisited. Granul. Matter **9**, 465–469 (2007)

Thornton, C.: Coefficient of restitution for collinear collisions of elastic-perfectly plastic spheres. Trans. ASME J. Appl. Mech. **64**, 383–386 (1997)

Thornton, C., Cummins, S.J., Cleary, P.W.: An investigation of the comparative behaviour of alternative contact force models during elastic collisions. Powder Technol. **210**, 189–197 (2011)

Thornton, C., Cummins, S.J., Cleary, P.W.: An investigation of the comparative behaviour of alternative contact force models during inelastic collisions. Powder Technol. **233**, 30–46 (2013)

Tsuji, Y., Tanaka, T., Ishida, T.: Lagrangian numerical simulation of plug flow of cohesionless particles in a horizontal pipe. Powder Technol. **71**, 239–250 (1992)

Walton, O.R.: Application of molecular dynamics to macroscopic particles. Int. J. Eng. Sci. **22**, 1097–1107 (1983)

Walton, O.R.: Numerical simulation of inclined chute flows of monodisperse, inelastic frictional spheres. Mech. Mater. **16**, 239–247 (1993)

Walton, O.R., Braun, R.I.: Viscosity, granular temperature, and stress calculations for shearing assemblies of inelastic, frictional disks. J. Rheol. **30**, 949–980 (1986)

Chapter 5
Particle Impact

Abstract Understanding the impact between two particles is of fundamental importance in numerous engineering applications and scientific studies. A binary collision may appear to be a very simple problem but, in fact, it is a very complex event. This is due to the short duration and the high localised stresses generated that, in most cases, result in both frictional and plastic dissipation. In addition, if rigid body sliding does not occur throughout the impact, then local elastic deformation of the two bodies becomes significant. This chapter examines both normal and oblique impacts of a sphere with a target wall and considers the effects of elasticity, plastic dissipation, surface energy and initial spin.

Understanding the impact between two particles is of fundamental importance in numerous engineering applications and scientific studies. A binary collision may appear to be a very simple problem but, in fact, it is a very complex event. This is due to the short duration and the high localised stresses generated that, in most cases, result in both frictional and plastic dissipation. In addition, if rigid body sliding does not occur throughout the impact, then local elastic deformation of the two bodies becomes significant. In the powder processing industries and in nature, the problem is further complicated due to the fact that the particles are non-spherical and the outcome of an impact event depends on the particle shape and orientation. For simplicity, in this chapter, only spherical particles will be considered.

5.1 Normal Impact

The original pioneering work on impact of spheres is due to Hertz (1896). Following directly from his theory of elastic contact, Hertz analysed the impact of frictionless elastic bodies. The assumption made in this quasi-static theory is that the deformation is restricted to the vicinity of the contact area. The theory also assumes that the energy loss due to elastic wave propagation can be ignored and the total mass of each body moves at any instant with the velocity of its centre of mass.

C. Thornton, *Granular Dynamics, Contact Mechanics and Particle System Simulations*, Particle Technology Series 24, DOI 10.1007/978-3-319-18711-2_5

A particular consequence of the quasi-static nature of the model is that the elastic energy of the two bodies acquired during the collision is entirely reversible, the magnitudes of the impact and rebound velocities are identical and the normal coefficient of restitution e_n is unity.

5.1.1 Elastic Impact

From the theory of Hertz we can rewrite Eq. (3.13) as

$$m^* \frac{d^2\alpha}{dt^2} = -\frac{4}{3}E^* {R^*}^{1/2} \alpha^{3/2} \tag{5.1}$$

Since, for a sphere of mass m and radius R impacting a massive target wall, $m^* = m$ and $R^* = R$. Integrating with respect to α gives

$$\frac{m}{2}\left[V_i^2 - \left(\frac{d\alpha}{dt}\right)^2\right] = \frac{8}{15}E^* R^{1/2} \alpha^{5/2} \tag{5.2}$$

where V_i is the velocity of approach at $t = 0$. When the displacement reaches the maximum value there is no relative motion between the two bodies and therefore

$$\alpha_{max} = \left(\frac{15mV_i^2}{16E^* R^{1/2}}\right)^{2/5} \tag{5.3}$$

The duration of the impact t_c is given (Raman 1920) by

$$t_c = 2.94\frac{\alpha_{max}}{V_i} \tag{5.4}$$

thus

$$t_c = 2.865\left(\frac{m^2}{RE^{*2}V_i}\right)^{1/5} \tag{5.5}$$

The validity of the Hertz theory was demonstrated experimentally by Andrews (1930), who investigated the impact of two equal spheres of soft metal at low impact velocities and confirmed that $t_c \propto V_i^{-1/5}$ and the coefficient of restitution is very close to unity. Energy loss due to elastic wave propagation during an elastic impact was analysed by Hunter (1957) who showed that, for a steel ball impacting a large block of steel or glass, less than one per cent of the initial kinetic energy was converted into elastic waves. At higher velocities, energy dissipation occurs due to plastic deformation or fracture. It was shown by Hutchings (1979) that only a few per cent of the initial kinetic energy is dissipated by stress waves during plastic impacts.

5.1.2 Effect of Plastic Dissipation

Consider the normal impact of two elastic-perfectly plastic spheres. If the relative impact velocity V_i is just large enough to initiate yield in one of the spheres then, using Eqs. (3.12) and (3.13), we may write

$$\frac{1}{2}m^* V_y^2 = \int_0^{a_y} F_n d\alpha = \frac{8E^* a_y^5}{15R^{*2}} \tag{5.6}$$

where V_y, defined as the yield velocity, is the relative impact velocity below which the interaction is assumed to be elastic, a_y is the contact radius when yield occurs and m^* is related to the two particle masses by $1/m^* = 1/m_1 + 1/m_2$. Rearranging Eq. (5.6),

$$a_y = \left(\frac{15R^{*2} m^* V_y^2}{16E^*}\right)^{1/5} \tag{5.7}$$

Combining Eqs. (3.12), (3.48) and (3.50), the limiting contact pressure p_y may be written as

$$p_y = \frac{2E^* a_y}{\pi R^*} \tag{5.8}$$

Then, from Eqs. (5.7) and (5.8) the yield velocity is given by

$$V_y = \left(\frac{\pi}{2E^*}\right)^2 \left(\frac{8\pi R^{*3}}{15m^*}\right)^{1/2} p_y^{5/2} = 3.194\left(\frac{p_y^5 R^{*3}}{E^{*4} m^*}\right)^{1/2} \tag{5.9}$$

In the case of a sphere of density ρ impacting a massive plane target wall, $R^* = R$, $m^* = m$ and Eq. (5.9) reduces to

$$V_y = \left(\frac{\pi}{2E^*}\right)^2 \left(\frac{2}{5\rho}\right)^{1/2} p_y^{5/2} = 1.56\left(\frac{p_y^5}{E^{*4}\rho}\right)^{1/2} \tag{5.10}$$

which was originally obtained by Davies (1949).

Thornton (1997) considered the normal impact of two elastic-perfectly plastic spheres, see Sect. 3.2.1 for details of the contact mechanics formulation, and derived the following analytical expression for the normal coefficient of restitution.

$$e_n = \left(\frac{6\sqrt{3}}{5}\right)^{1/2} \left[1 - \frac{1}{6}\left(\frac{V_y}{V_i}\right)^2\right]^{1/2} \left[\frac{\left(\frac{V_y}{V_i}\right)}{\left(\frac{V_y}{V_i}\right) + 2\sqrt{\frac{6}{5} - \frac{1}{5}\left(\frac{V_y}{V_i}\right)^2}}\right]^{1/4} \tag{5.11}$$

which satisfies the condition $e_n = 1.0$ when $V_i = V_y$ and, at high impact velocities when $(V_y/V_i)^2 \to 0$

$$e_n = \left(\frac{6\sqrt{3}}{5}\right)^{1/2} \left(\frac{V_y}{V_y + \frac{2\sqrt{6}}{\sqrt{5}} V_i}\right)^{1/4} \tag{5.12}$$

Then, taking $V_i \gg V_y$ we obtain

$$e_n = \left(\frac{6\sqrt{3}}{5}\right)^{1/2} \left(\frac{\sqrt{5}}{2\sqrt{6}}\right)^{1/4} \left(\frac{V_y}{V_i}\right)^{1/4} = 1.185 \left(\frac{V_y}{V_i}\right)^{1/4} \tag{5.13}$$

For the case of a sphere impacting a plane surface we may substitute for V_y using Eq. (5.10) to obtain

$$e_n = 1.324 \left(\frac{p_y^5}{E^{*4} \rho}\right)^{1/8} V_i^{-1/4} \tag{5.14}$$

A similar expression was provided by Johnson (1985) except that the prefactor was 1.72 as a result of assuming that the plastic normal stiffness was twice that assumed by Thornton (1997).

Generally, there are four deformation regimes depending on the impact velocity or amount of indentation, namely the elastic, elastoplastic, plastic and finite plastic deformation regimes, Mesarovic and Fleck (1999). The initial deformation is elastic with the contact force increasing at an increasing rate, as defined by Hertz (Johnson 1985). This is followed by a complex elastoplastic regime.

At the transition from elastic to elastoplastic behaviour the force-displacement curve deviates from the Hertzian curve and is tangent to the Hertzian curve at the point when initial yield occurs. Subsequently the contact force increases at an increasing rate due to strain hardening during which the effective yield stress increases at a decreasing rate. Li et al. (2002) developed a rather complicated theoretical model to incorporate strain hardening effects and Kharaz and Gorham (2000) presented results of normal impact experiments that illustrated the velocity dependency of the restitution coefficient in this regime.

At sufficiently high impact velocities, the behaviour is perfectly plastic and the contact stiffness is constant, as assumed by Thornton (1997). This is followed by a finite plastic deformation regime at very high impact velocities. In this regime the plastic deformation of the sphere extends beyond the contact area and the consequence is that the contact force increases at a decreasing rate, Mesarovic and Fleck (1999). From the results of their finite element analyses, Wu et al. (2003) showed that in this regime the exponent of the power law relationship between the coefficient of restitution and the impact velocity suddenly changed from −0.25 to −0.5. Interestingly, the same abrupt change in the velocity exponent was obtained by Schrapler et al. (2012) from their DEM simulations of agglomerate impact.

5.1.3 Effect of Surface Energy

For adhesive elastic spheres, as shown in Fig. 3.13, separation occurs at point D when $\alpha = -\alpha_f$ and hence the work required to break the contact W_s is given by the area under the curve for $0 > \alpha > -\alpha_f$. Hence

$$W_s = 7.09\left(\frac{\Gamma^5 R^{*4}}{E^{*2}}\right)^{1/3} \tag{5.15}$$

Neglecting energy losses due to elastic wave propagation we may write

$$\frac{1}{2}m^* V_i^2 - \frac{1}{2}m^* V_r^2 = W_s \tag{5.16}$$

If the rebound velocity $V_r = 0$ then the impact velocity $V_i = V_s$, the critical velocity below which sticking occurs and from Eqs. (5.15) and (5.16) we obtain the sticking criterion

$$V_s = \left(\frac{14.18}{m^*}\right)^{1/2}\left(\frac{\Gamma^5 R^{*4}}{E^{*2}}\right)^{1/6} \tag{5.17}$$

For a sphere impacting a flat surface, $R^* = R$ and $m^* = m$ leading to

$$V_s = 1.84\left[\frac{(\Gamma/R)^5}{\rho^3 E^{*2}}\right]^{1/6} \tag{5.18}$$

If $V_i > V_s$, then bounce occurs and we may rewrite Eq. (5.16) as

$$1 - \left(\frac{V_r}{V_i}\right)^2 = \left(\frac{V_s}{V_i}\right)^2 \tag{5.19}$$

from which the coefficient of restitution is defined by

$$e_n = \left[1 - \left(\frac{V_s}{V_i}\right)^2\right]^{1/2} \tag{5.20}$$

For $V_i > V_s$, the coefficient of restitution increases at a decreasing rate and, as shown by Thornton and Ning (1998), when the impact velocity is ten times higher than the critical sticking velocity the coefficient of restitution is 0.995.

For adhesive, elastic-plastic spheres, Thornton and Ning (1998) assumed that the work dissipated due to plastic deformation and the work dissipated due to adhesive rupture are additive, i.e.

$$\left(1 - e_n^2\right) = \left(1 - e_p^2\right) + \left(1 - e_a^2\right) \tag{5.21}$$

where e_p is the coefficient of restitution due to plastic deformation given by Eq. (5.11) and e_a is the coefficient of restitution due to adhesive rupture given by Eq. (5.20). This leads to the following equations.

$$e_n = 0 \quad \text{for} \quad V_i \le V_s \tag{5.22a}$$

$$e_n = \left[1 - \left(\frac{V_s}{Vi}\right)^2\right]^{1/2} \quad \text{for} \quad V_s \le V_i \le V_y \tag{5.22b}$$

and for $V_i \ge V_y$ the coefficient of restitution can be obtained from

$$e_n^2 = \left(\frac{6\sqrt{3}}{5}\right)\left[1 - \frac{1}{6}\left(\frac{V_y}{V_i}\right)^2\right]\left[\frac{\left(\frac{V_y}{V_i}\right)}{\left(\frac{V_y}{V_i}\right) + 2\sqrt{\frac{6}{5} - \frac{1}{5}\left(\frac{V_y}{V_i}\right)^2}}\right]^{1/2} - \left(\frac{V_s}{V_i}\right)^2 \tag{5.22c}$$

The solution to the above equations is illustrated in Fig. 5.1 by plotting the coefficient of restitution against the normalised velocity (V_i/V_y) for different ratios of (V_s/V_y).

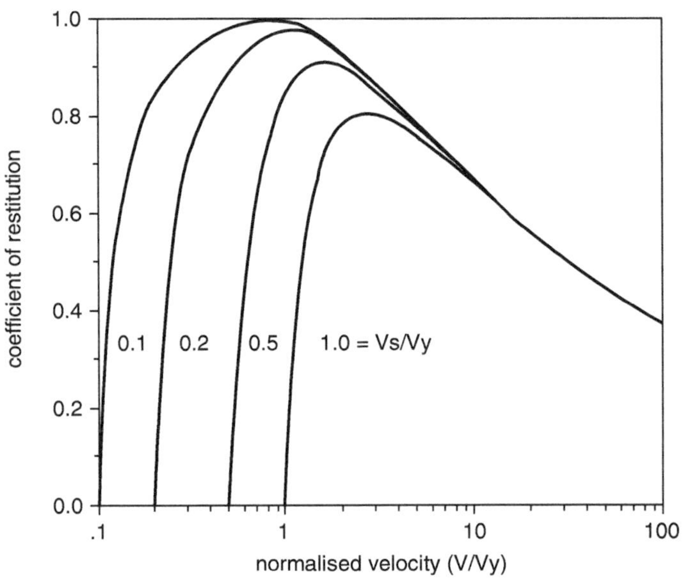

Fig. 5.1 Dependency of the coefficient of restitution on the yield velocity and the critical sticking velocity for adhesive elastic-plastic interactions

5.2 Oblique Impact

In particle system simulations normal impact is a rare event except when particles collide with wall boundaries. If two spheres collide then only if their velocities are such that the centres of both spheres would arrive at exactly the same point in space at exactly the same time will the collision be a normal impact. For simplicity we focus on oblique impacts of a sphere with a planar target wall.

The problem is illustrated diagrammatically in Fig. 5.2. The sphere approaches the wall with an initial translational velocity V_i at an impact angle θ_i. After interaction with the wall, the sphere rebounds at an angle θ_r with a rebound translational velocity V_r and a rebound angular velocity ω_r. Note that V_i and V_r are the velocities of the sphere centre. The corresponding tangential components of the surface velocities at the contact are denoted by v_{ti} and v_{tr} that are related to the translational velocities by

$$v_{ti} = V_{ti} + R\omega_i \qquad \text{and} \qquad v_{tr} = V_{tr} + R\omega_r \tag{5.23}$$

Note that, in the figure, Z and X indicate the positive axes and that anti-clockwise rotation is positive. Also, the sphere approaches the wall without any rotation, there is no gravity field, and the wall is considered to be massive in comparison with the sphere. The effect of initial rotation will be discussed later in the chapter.

Results of simulations of an elastic sphere impacting an elastic wall at different impact angles are shown in Figs. 5.3, 5.4 and 5.5. The contact force model used is

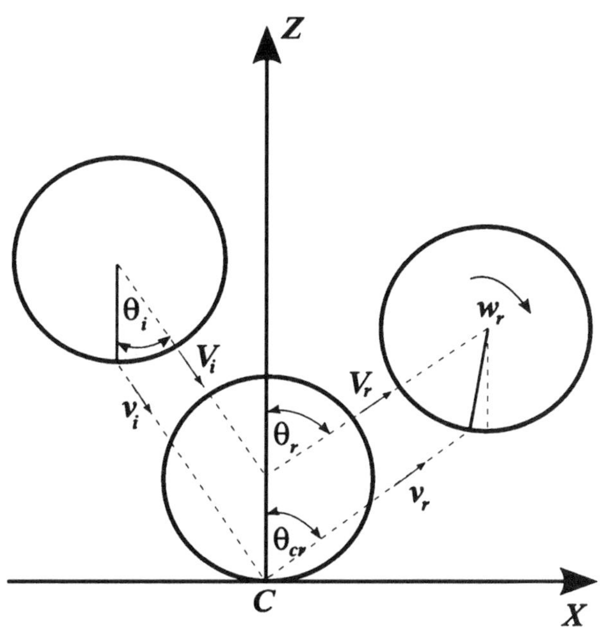

Fig. 5.2 Diagram of the oblique impact of a sphere with a plane surface

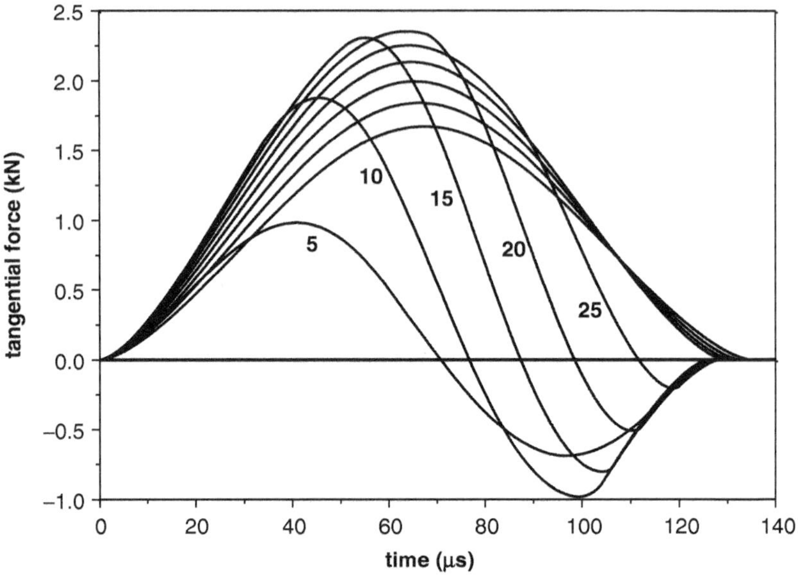

Fig. 5.3 Evolution of the tangential force for different impact angles

described in Sect. 3.1, i.e. Hertz (normal), Mindlin and Deresiewicz (tangential). The sphere and the wall had the same elastic properties and the coefficient of interface friction $\mu = 0.1$. For all impact angles the impact speed was 5.0 m/s.

Figure 5.3 shows the time evolution of the tangential force for impact angles in the range 5–45°. For impact angles ≥30° the tangential force never reverses direction because in this range the contact is sliding, $F_t = \mu F_n$, throughout the impact duration. Figure 5.4 shows the corresponding tangential force-displacement curves from which it can be seen that for very small impact angles, e.g. 5° and 10°, the displacement is negative at the end of the impact.

In Fig. 5.5 the tangential force is plotted against the normal force. When the impact angle is 5°, sliding does not occur ($F_t < \mu F_n$) until the very end of the impact. This behaviour applies if $tan\theta_i \leq \mu/\kappa$. For $\theta = 10°$ and 15°, sliding occurs from the start of the impact and continues until the decelerating tangential translational motion of the sphere and the accelerating particle rotation induced by the tangential force combine to reduce the tangential force increment to $\Delta F_t < \mu \Delta F_n$. At this point, in the context of the theory of Mindlin and Deresiewicz (1953), Sect. 3.1.2, a stick region starts to grow from the centre of the contact area. Subsequently, as the resultant force rotates, the tangential force reduces, reverses in direction and finally, towards the end of the impact, sliding reoccurs. For $\theta = 20°$ and 25°, the initial sliding condition continues into the restitution stage, when the normal force is reducing, before the $|\Delta F_t| < \mu \Delta F_n$ condition occurs. This is then followed by rotation of the resultant force and finally sliding occurs towards the end of the impact. For $\theta \geq 30°$, the sliding condition, $F_t = \mu F_n$, applies throughout the impact with no reversal of the tangential force direction.

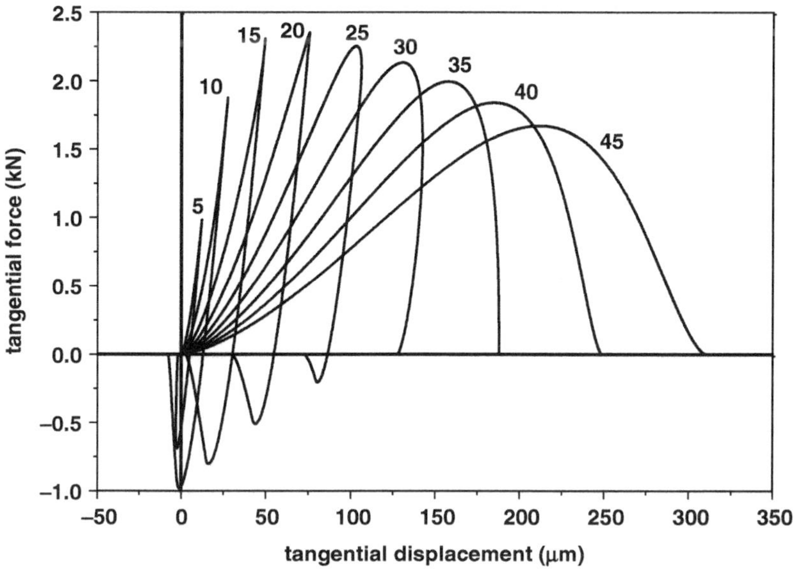

Fig. 5.4 Tangential force-displacement curves for different impact angles

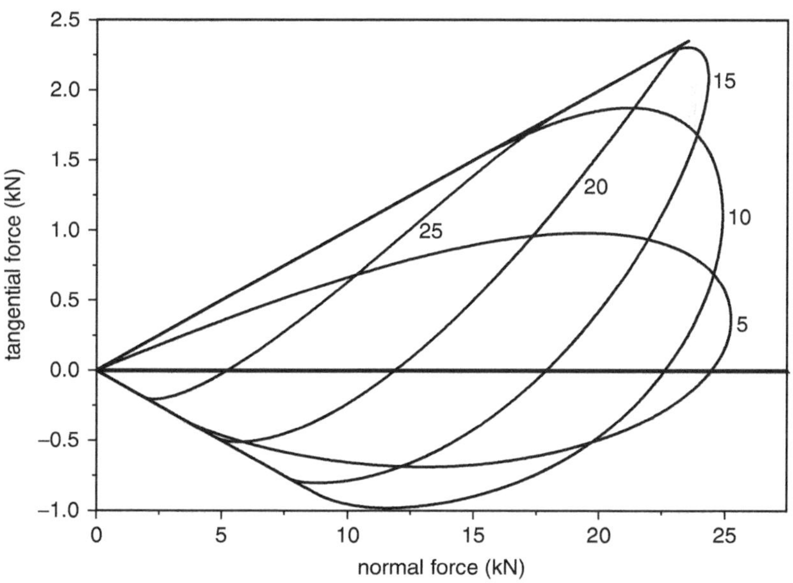

Fig. 5.5 Evolution of the tangential force and the normal force for different impact angles

Of primary importance in particle collisions is to be able to define the rebound kinematics. At the end of the impact, we need to know the direction and spin of the rebounding sphere because this will affect the next collision. In other words, we need to predict the complete rebound kinematics.

The simplest theoretical approach to the oblique impact of a sphere with a target wall is that of rigid dynamics (Goldsmith 1960; Brach 1991; Stronge 2000). However, the approach is only valid if the impact angle is sufficiently large that sliding occurs throughout the impact duration. At smaller impact angles, the theory predicts that the tangential surface velocity is zero at the end of the impact event. That this is not the case was demonstrated by Maw et al. (1976, 1981) who showed that the tangential surface velocity reverses its direction due to the tangential elastic compliance. Nevertheless, rigid body dynamics does provide appropriate dimensionless groups that characterise the kinematic behaviour of oblique impacts.

5.2.1 Rigid Body Dynamics

In rigid body dynamics, rebound velocities are related to the impact velocities by empirical coefficients of restitution in the normal and tangential direction, defined by

$$e_n = \frac{-V_{nr}}{V_{ni}} \quad \text{and} \quad e_t = \frac{V_{tr}}{V_{ti}} \tag{5.24}$$

where the subscripts n and t indicate the normal and tangential components respectively. The correlation between the tangential and normal interactions during an impact, is characterised by an impulse ratio defined as

$$f = \frac{P_t}{P_n} = \frac{\int F_t dt}{\int F_n dt} \tag{5.25}$$

where P_n and P_t are the normal and tangential impulses respectively. If sliding occurs throughout the impact duration then the impulse ratio is equal to the interface friction coefficient $f = \mu$, otherwise $f < \mu$. According to Newton's second law, the normal and tangential impulses can be expressed in terms of the impact and rebound velocities as

$$P_n = m(V_{nr} - V_{ni}) \quad \text{and} \quad P_t = m(V_{tr} - V_{ti}) \tag{5.26}$$

where m is the mass of the sphere. Combining Eqs. (5.24), (5.25) and (5.26) we obtain

$$e_t = 1 - \frac{f(1 + e_n)}{tan\,\theta_i} \tag{5.27}$$

By analogy, we can also define a rotational 'impulse'

$$P_\omega = I(\omega_r - \omega_i) \tag{5.28}$$

where I is the moment of inertia of the sphere and ω_i and ω_r are the initial and rebound angular velocities of the sphere. According to the conservation of angular momentum about the contact point

$$P_\omega = RP_t \tag{5.29}$$

where R is the radius of the sphere. From Eqs. (5.26), (5.28) and (5.29) we obtain

$$\omega_r - \omega_i = -\frac{mR}{I}(V_{ti} - V_{tr}) \tag{5.30}$$

For a solid sphere, $I = 2mR^2/5$. Hence,

$$\omega_r - \omega_i = -\frac{5(V_{ti} - V_{tr})}{2R} = -\frac{5V_{ti}(1 - e_t)}{2R} \tag{5.31}$$

or, using Eq. (5.27),

$$\omega_r - \omega_i = -\frac{5f(1 + e_n)V_{ni}}{2R} \tag{5.32}$$

The tangential component of the rebound surface velocity at the contact patch v_{tr} is obtained from Eqs. (5.23) and (5.32)

$$v_{tr} = V_{tr} + R\omega_i - \frac{5}{2}f(1 + e_n)V_{ni} \tag{5.33}$$

Combining Eqs. (5.24), (5.27) and (5.33) we then obtain

$$v_{tr} - v_{ti} = -\frac{7}{2}f(1 + e_n)V_{ni} \tag{5.34}$$

and, finally, the rebound angle θ_r is obtained from

$$tan\,\theta_r = \frac{V_{tr}}{V_{nr}} = -\frac{e_t}{e_n}tan\,\theta_i \tag{5.35}$$

5.2.2 Dimensionless Groups

If sliding occurs throughout the impact duration then $f = \mu$ and the following three equations are obtained to define the complete rebound kinematics

$$v_{tr} = v_{ti} - \frac{7}{2}(1 + e_n)\mu V_{ni} \tag{5.36}$$

$$\omega_r = \omega_i - \frac{5(1 + e_n)}{2R}\mu V_{ni} \tag{5.37}$$

$$e_t = \frac{V_{tr}}{V_{ti}} = 1 - \frac{\mu(1 + e_n)}{\tan\theta_i} \tag{5.38}$$

Rearranging Eqs. (5.36) and (5.37)

$$\frac{2v_{tr}}{(1 + e_n)\mu V_{ni}} = \frac{2v_{ti}}{(1 + e_n)\mu V_{ni}} - 7 \tag{5.39}$$

$$\frac{2R(\omega_r - \omega_i)}{(1 + e_n)\mu V_{ni}} = -5 \tag{5.40}$$

from which we identify three dimensionless groups, namely

$$\frac{2v_{ti}}{(1 + e_n)\mu V_{ni}}, \quad \frac{2v_{tr}}{(1 + e_n)\mu V_{ni}} \quad \text{and} \quad \frac{2R(\omega_r - \omega_i)}{(1 + e_n)\mu V_{ni}} \tag{5.41}$$

which, in the case of no initial rotational velocity, reduce to

$$\frac{2\tan\theta_i}{(1 + e_n)\mu}, \quad \frac{2v_{tr}}{(1 + e_n)\mu V_{ni}} \quad \text{and} \quad \frac{2R\omega_r}{(1 + e_n)\mu V_{ni}} \tag{5.42}$$

Maw et al. (1976, 1981) suggested, for elastic impacts between similar bodies, alternative dimensionless groups to normalise the data that can be written as

$$\frac{\kappa \tan\theta_i}{\mu} \quad and \quad \frac{\kappa v_{tr}}{\mu V_{ni}} \tag{5.43}$$

However, it was demonstrated by Thornton (2009) that, for different values of κ, the data was not normalised, even in the sliding regime.

5.2.3 Effect of Elasticity

The elastic moduli affect the magnitude of the contact forces and the contact duration. The rebound kinematics, defined in terms of the above dimensionless groups, are sensitive to the ratio of tangential to normal elastic stiffnesses, κ,

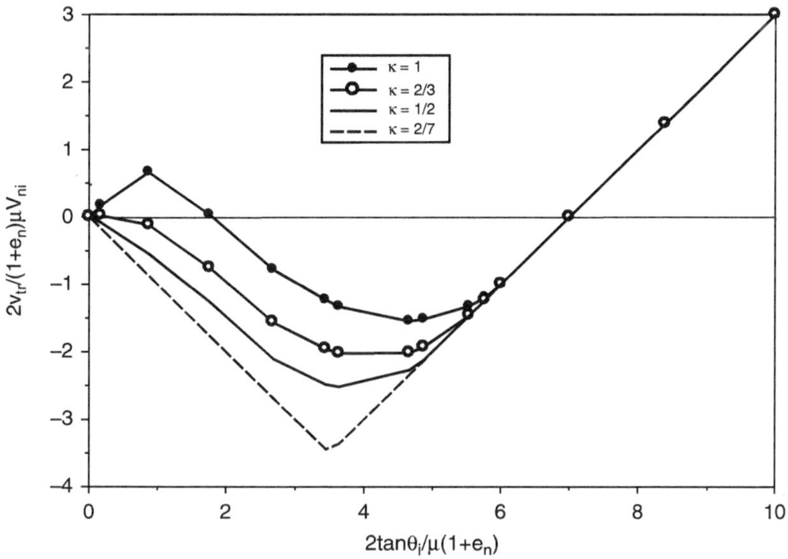

Fig. 5.6 Effect of κ on the normalised rebound surface velocity

because this parameter controls the ratio of the frequencies of the tangential and normal force oscillations. For dissimilar materials κ depends on the elastic moduli and the Poisson's ratios of the two colliding bodies, see Eq. (4.16), but if the elastic properties of the two bodies are the same then κ only depends on ν, see Eq. (4.17).

The rebound kinematics, expressed in terms of the dimensionless groups, are shown in Figs. 5.6 and 5.7 and the tangential coefficient of restitution e_t is shown in Fig. 5.8, for different values of κ. The frequency of the tangential force oscillation is exactly twice that of the normal force oscillation when $\kappa = 1$. When $\kappa = 2/7$, the frequencies of the tangential and normal force oscillations are equal, which is physically unrealistic since this means that the obliquity of the resultant contact force is constant throughout the collision with $F_t = 2/7 tan\, \theta_i F_n$ or, if sliding occurs throughout the impact, $F_t = \mu F_n$. As pointed out in Sect. 4.2, see Eq. (4.17), physically realistic values of κ are limited to the range $1 \geq \kappa \geq 2/3$ since Poisson's ratio is limited to the range $0 \leq \nu \leq 0.5$. If sliding occurs throughout the impact then the data points, in Fig. 5.6, lie on the inclined straight line defined by Eq. (5.39) and, in Fig. 5.7, lie on the horizontal line defined by Eq. (5.40). The figures show that sliding throughout the impact depends on the value of κ. Rigid body dynamics predicts that sliding occurs throughout the impact if $tan\theta_i \geq 7\mu$. However, for elastic impacts, sliding occurs throughout the impact duration at smaller impact angles when the following condition is satisfied.

$$\frac{tan\, \theta_i}{\mu} \geq 7 - 1/\kappa \qquad (5.44)$$

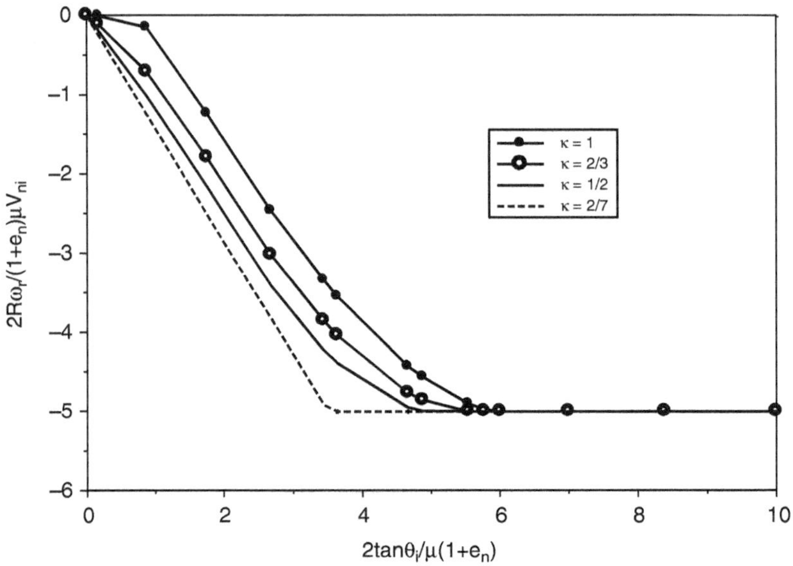

Fig. 5.7 Effect of κ on the normalised rebound angular velocity

If sliding does not occur throughout the impact then, for no initial spin, the tangential coefficient of restitution is defined by

$$e_t = \frac{5}{7} + \frac{2v_{tr}}{7V_{ti}} \tag{5.45}$$

In this case, if $\kappa = 2/7$ then $e_t = 3/7$ and rigid body dynamics ($v_{tr} = 0$) predicts $e_t = 5/7$. It is also worth noting that, since all the data sets in Fig. 5.6 pass through the origin, the tangential coefficient of restitution is indeterminate when the impact angle is zero.

Comparisons of the rebound kinematics obtained using linear and non-linear spring models (no dashpots) with the results obtained using the Hertz, Mindlin and Dereseiwicz model are provided by Thornton et al. (2011).

5.2.4 *Effect of Plastic Dissipation*

In this section we examine the effect of plastic dissipation on the rebound characteristics using the contact force models of Thornton (1997) for the normal interaction and Mindlin and Deresiewicz (1953) for the tangential interaction, see Sect. 3.2 for details. For comparisons with the results of other contact force models the reader is directed to Thornton et al. (2013).

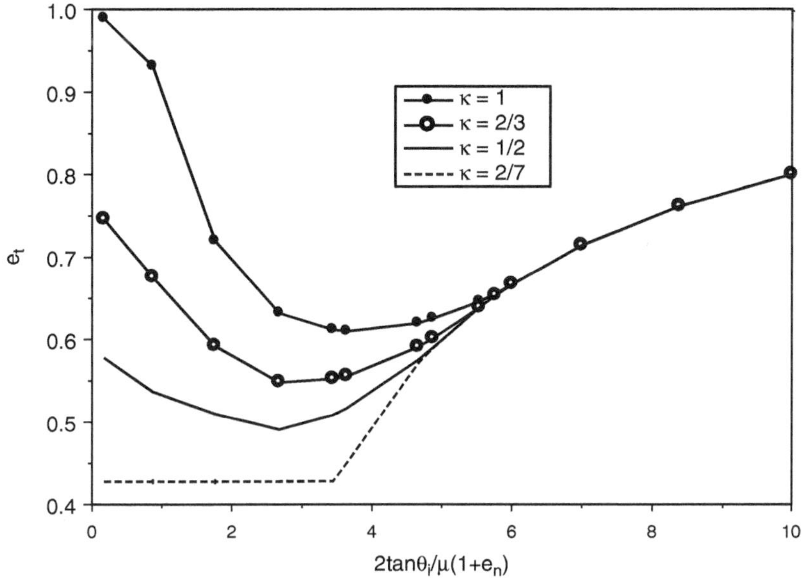

Fig. 5.8 Effect of κ on the tangential coefficient of restitution

Consider the oblique impact of a sphere with a target wall, as shown in Fig. 5.2. For the results presented in this section, the physical properties of the sphere are radius $R = 25$ mm, density $\rho = 2650$ kg/m^3 and hence a mass $m = 0.1734$ kg. The elastic properties of both the sphere and the wall are $E = 70$ GPa, $\nu = 0.3$, therefore $\kappa = 0.8235$, and the interface friction $\mu = 0.1$. A constant normal component of the impact speed $V_{ni} = 5$ m/s, and hence a constant normal coefficient of restitution, is used for all impact angles. For a selected value of e_n the yield velocity V_y is obtained from Eq. (5.11) and substituted into Eq. (5.10) to obtain the required value for the limiting contact pressure p_y.

In Fig. 5.9, the tangential force evolution for different impact angles is shown for values of the normal coefficient of restitution $e_n = 1.0$, 0.75, 0.5 and 0.25. It can be seen that the frequencies of the tangential force oscillations change when the normal coefficient of restitution is changed. The figure shows that, for $e_n = 0.5$, there is a second reversal of the tangential force when the impact angle is 5°. For $e_n = 0.25$, this is also true when the impact angle is 10° and when the impact angle is 5° there is a third reversal. Multiple reversals of tangential force have been observed experimentally by Cross (2002) albeit for a hollow basketball impacting at an angle of 24°. Becker et al. (2008) showed that, using linear springs, by arbitrarily increasing the tangential to normal stiffness ratio to values of $\kappa \gg 1$ they obtained multiple reversals of the relative tangential surface displacement, and hence the tangential force, for small impact angles. With the Thornton, Mindlin and Deresiewicz model, ratios of tangential to normal stiffness greater than unity arise

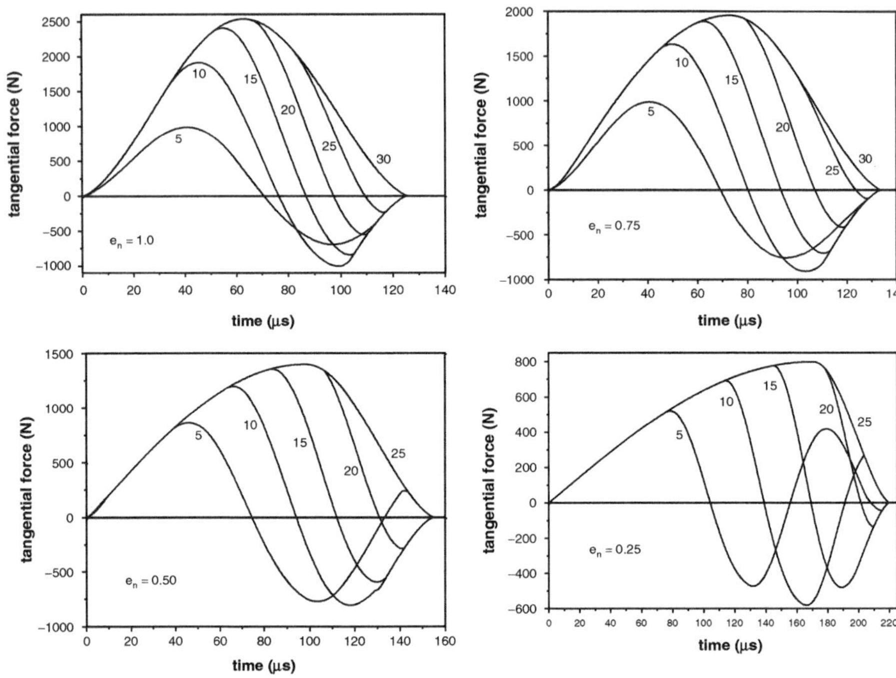

Fig. 5.9 Evolution of tangential force for different normal coefficients of restitution

due to the decrease in normal loading stiffness and the increase in tangential loading stiffness when e_n is decreased.

Figure 5.9 also shows that the value of e_n affects the range of impact angles for which sliding occurs throughout the impact duration and the range of small impact angles for which sliding does not occur until towards the end of the impact. Sliding occurs throughout the impact if the following inequality is true

$$tan\,\theta_i \geq \frac{\mu(1+e_n)}{2}\left(7-\frac{e_n}{\kappa}\right) \tag{5.46}$$

and sliding only occurs at the end of the impact if

$$tan\,\theta_i \leq \frac{\mu(1+e_n)e_n}{2\kappa} \tag{5.47}$$

The effect of the plastic dissipation on the rebound kinematics is shown in Figs. 5.10 and 5.11. Figure 5.10 shows that the range of small impact angles for which the rebound tangential surface velocity is positive increases as the normal coefficient of restitution reduces to $e_n = 0.5$ and that there is a reduction in the maximum negative tangential surface velocity. When $e_n = 0.25$, additional fluctuations appear and as $e_n \rightarrow 0$, $v_{tr} \rightarrow 0$, coinciding with the rigid body dynamics

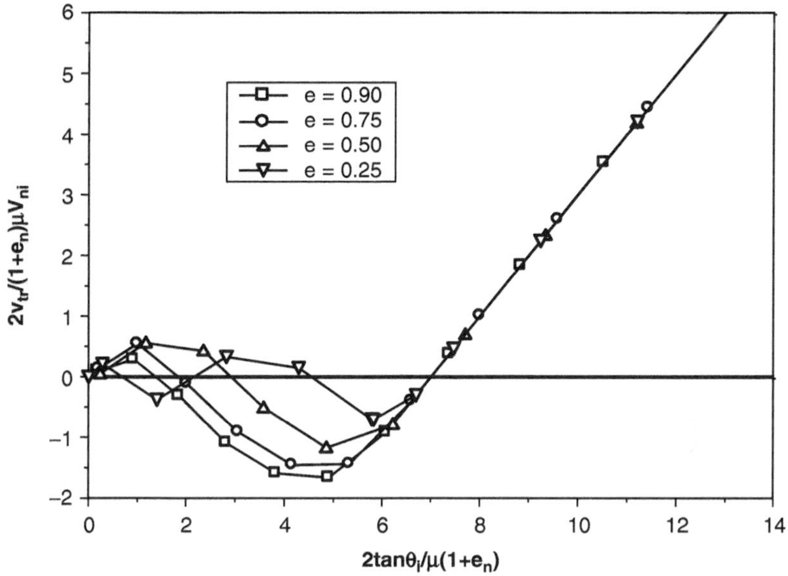

Fig. 5.10 Effect of e_n on the tangential surface velocity v_{tr} at rebound

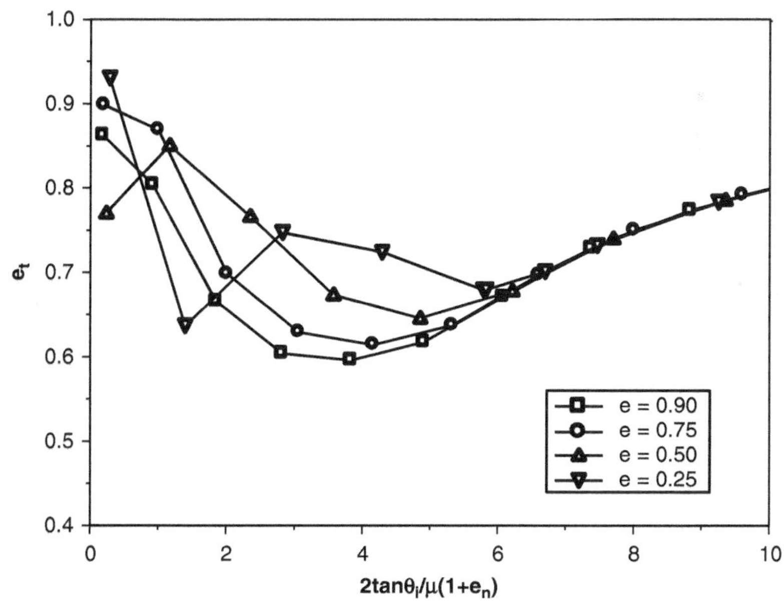

Fig. 5.11 Effect of e_n on the tangential coefficient of restitution

solution for impacts during which sliding does not occur throughout the impact. The fluctuations in the rebound tangential surface velocity are due to the way in which the normal coefficient of restitution changes the frequencies of the tangential force oscillations, as shown in Fig. 5.9. It was demonstrated by Thornton et al. (2013) that the sign and magnitude of the rebound tangential surface velocity corresponds to the slope of the tangential force-time curve that would have occurred at the end of the impact, if sliding had not occurred.

Because of Eq. (5.45), for impacts during which sliding does not occur throughout the impact, the fluctuations in the rebound tangential surface velocity shown in Fig. 5.10 are reflected in oscillations in the tangential coefficient of restitution, as shown in Fig. 5.11.

5.2.5 *Effect of Initial Spin*

In this chapter, for simplicity, we have only considered impacts with no initial spin. This is quite unrealistic since, even in this case, the spin imparted during the collision will become the initial spin for the next collision. In the previous subsections the normalised rebound kinematics have been plotted against the normalised impact angle. However, the contact force reactions are functions of the relative surface velocities.

If there is no initial spin then the tangential surface velocity is $v_x = V_x$ and the impact angle is defined by $tan\theta = V_x/V_y$, as shown in Fig. 5.12a. If there is an initial in-plane spin the tangential surface velocity is $v_x = V_x + R\omega_z$ where ω_z is the rotational velocity about the z axis, as shown in Fig. 5.12b.

The contact reaction thinks that the sphere is coming in at a different angle defined by

$$tan\,\theta' = \frac{v_x}{V_y} = \frac{V_x + R\omega_z}{V_y} \tag{5.48}$$

where θ' may be referred to as the ‘effective impact angle’, see Thornton (2009). In this chapter the rebound kinematics have been presented in terms of the dimensionless groups given by Eq. (5.42). If there is an initial in-plane spin then exactly the same curves are obtained by using the dimensionless groups defined by Eq. (5.41).

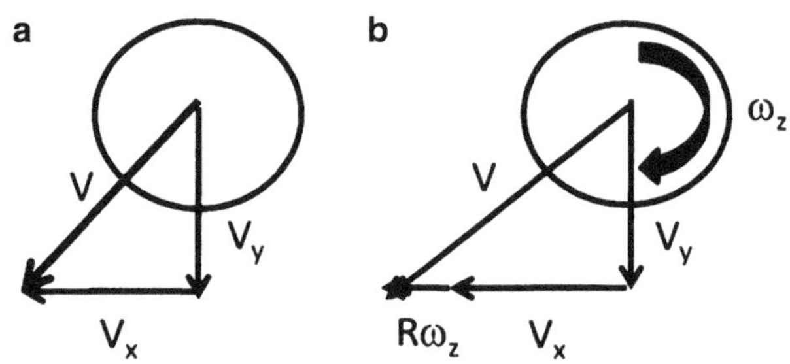

Fig. 5.12 Impact with a horizontal wall (**a**) no initial spin (**b**) with initial spin

If there is an additional out-of-plane initial spin component about the x-direction then the resultant direction of the tangential surface velocity v is inclined to the x-direction by an angle β as shown in Fig. 5.13a. If we consider that the tangential surface velocity components shown in Fig. 5.13a are the initial values at the start of the impact then, since $v_z = -R\omega_x$, $v_x = V_x + R\omega_z$ and $v_z/v_x = tan\beta$, the ratio of the components of the tangential force T at the end of the first time step is $T_z/T_x = tan\beta$ since $T = \Delta T = k_t v \Delta t$.

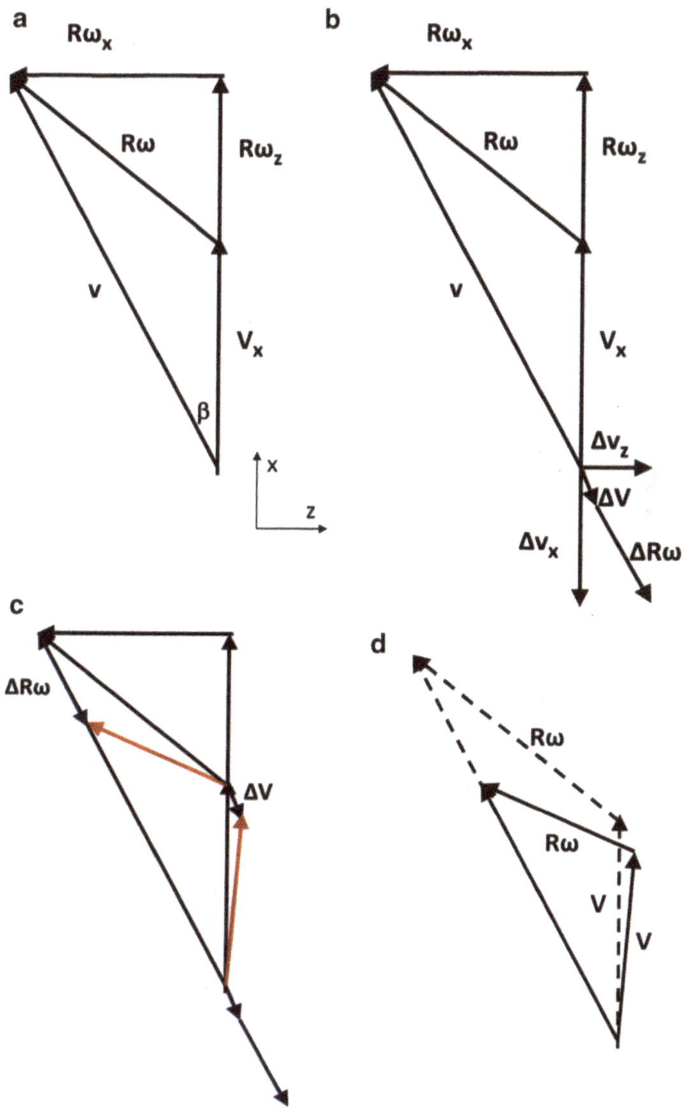

Fig. 5.13 Tangential surface velocities for out-of plane spin

The tangential force will then decelerate both the translational and rotational velocities of the sphere and will create new velocity increments given by

$$\Delta V_z = (T_z/m)\Delta t \ \text{ and } \ \Delta V_x = (T_x/m)\Delta t \tag{5.49}$$

$$\Delta\omega_z = (5T_z/2m)\Delta t \ \text{ and } \ \Delta\omega_x = (5T_x/2m)\Delta t \tag{5.50}$$

from which $\Delta V_z/\Delta V_x = \Delta R\omega_x/\Delta R_z = T_z/T_x$ and hence $\Delta v_z/\Delta v_x = T_z/T_x = tan\beta$.

The incremental velocity components are shown in Fig. 5.13b. The resultant velocity increments ΔV and $\Delta R\omega$ are relocated in Fig. 5.13c to show the new velocity vectors, in red, which indicate that during the impact the trajectory of the sphere centre changes direction and the direction of the plane of spin rotates. This is also evident in Fig. 5.13d which shows the old and new surface velocity components as dashed arrows and solid arrows respectively. However, the plane of spin remains orthogonal to the wall. It is important to note that the direction of the total surface velocity and the direction of the tangential force do not change during the impact. Consequently, if the initial tangential surface velocity v shown in Fig. 5.13a is used to define the 'effective impact angle' then, exactly the same normalised rebound surface velocities will be obtained. Predictions of the change in angular velocity and tangential coefficient of restitution are more complex and have not been fully examined.

The general impact case is when the initial plane of spin is not orthogonal to the wall. In this case, there is a component of rotation about the contact normal direction, ω_y, and this will create an additional reaction, i.e. a contact moment, as discussed in Sect. 3.1.3.

5.2.6 *Effect of Surface Energy*

There has been very little research on oblique impacts of adhesive particles. For non-adhesive elastic spheres, Thornton and Yin (1991) showed that for very small impact angles sliding did not occur until the end of the impact. For adhesive elastic spheres they found that, for the same range of small impact angles, the sphere rebounded back along the initial impact direction without any imparted spin, i.e. the tangential coefficient of restitution $e_t = -1$. For larger impact angles the rebound kinematics were very similar but not exactly the same as obtained for non-adhesive elastic spheres.

References

Andrews, J.P.: Experiment on impact. Proc. Phys. Soc. **43**, 8–17 (1930)

Becker, V., Schwager, T., Pöschel, T.: Coefficient of tangential restitution for the linear dashpot model. Phys. Rev. E **77**, 011304 (2008) (12 pages)

Brach, R.M.: Mechanical Impact Dynamics – Rigid Body Collisions. Wiley Interscience, New York (1991)

Cross, R.: Grip-slip behaviour of a bouncing ball. Am. J. Phys. **70**, 1093–1102 (2002)

Davies, R.M.: The determinations of static and dynamic yield stresses using a steel ball. Proc. Roy. Soc. London **A197**, 416–432 (1949)

Goldsmith, W.: Impact. Arnold, London (1960)

Hertz, H.: In: Jones, D.E., Schott, G.A. (eds.) Miscellaneous Papers by H. Hertz. MacMillan and Co., London (1896)

Hunter, S.C.: Energy absorbed by elastic waves during impact. J. Mech. Phys. Solids **5**, 162–171 (1957)

Hutchings, I.M.: Energy absorbed by elastic waves during plastic impact. J. Phys. D. Appl. Phys. **12**, 1819–1824 (1979)

Johnson, K.L.: Contact Mechanics. Cambridge University Press, Cambridge (1985)

Kharaz, A.H., Gorham, D.A.: A study of the restitution coefficient in elastic-plastic impact. Philos. Mag. Lett. **80**, 549–559 (2000)

Li, L.Y., Wu, C.Y., Thornton, C.: A theoretical model for the contact of elastoplastic bodies. Proc. IME. J. Mech. Eng. Sci. **216**, 421–431 (2002)

Maw, N., Barber, J.R., Fawcett, J.N.: The oblique impact of elastic spheres. Wear **38**, 101–114 (1976)

Maw, N., Barber, J.R., Fawcett, J.N.: The role of elastic tangential compliance in oblique impact. ASME J. Lubr. Technol. **103**, 74–80 (1981)

Mesarovic, S.D., Fleck, N.A.: Spherical indentation of elastic-plastic solids. Proc. Roy. Soc. London **A455**, 2707–2728 (1999)

Mindlin, R.D., Deresiewicz, H.: Elastic spheres in contact under varying oblique forces. Trans. ASME J. Appl. Mech. **20**, 327–344 (1953)

Raman, C.V.: On some applications of Hertz's theory of impact. Phys. Rev. **15**, 277–284 (1920)

Schrapler, R., Blum, J., Seizinger, A., Kley, W.: The physics of protoplanetesimal dust agglomerates. VII. The low-velocity collision behaviour of large dust agglomerates. Astrophys. J. **758**, 35 (2012). 9pp

Stronge, W.J.: Impact Mechanics. Cambridge University Press, Cambridge (2000)

Thornton, C.: Coefficient of restitution for collinear collisions of elastic-perfectly plastic spheres. Trans. ASME J. Appl. Mech. **64**, 383–386 (1997)

Thornton, C.: A note on the effect of initial particle spin on the rebound behaviour of oblique particle impacts. Powder Technol. **192**, 152–156 (2009)

Thornton, C., Ning, Z.: A theoretical model for the stick/bounce behaviour of adhesive, elasticplastic spheres. Powder Technol. **99**, 154–162 (1998)

Thornton, C., Yin, K.K.: Impact of elastic spheres with and without adhesion. Powder Technol. **65**, 153–166 (1991)

Thornton, C., Cummins, S.J., Cleary, P.W.: An investigation of the comparative behaviour of alternative contact force models during elastic collisions. Powder Technol. **210**, 189–197 (2011)

Thornton, C., Cummins, S.J., Cleary, P.W.: An investigation of the comparative behaviour of alternative contact force models during inelastic collisions. Powder Technol. **233**, 30–46 (2013)

Wu, C.Y., Li, L.Y., Thornton, C.: Rebound behaviour of spheres for plastic impacts. Int. J. Impact Eng. **28**, 929–946 (2003)

Chapter 6
Agglomerate Impacts

Abstract The storage, transport, handling and processing of particulate materials constitutes a significant part of the operations in most chemical, pharmaceutical and allied industries. These particulate materials are frequently in the form of powders which are, themselves, agglomerations of much smaller sized primary particles. A common problem inherent in the handling of powders is the degradation resulting from attrition and/or fragmentation of the agglomerates as they collide with each other and with the process equipment. In this chapter we focus on agglomerate breakage. It is shown that, for spherical agglomerates, dense systems fracture or shatter depending on the impact velocity. In contrast, loose systems disintegrate to a degree that depends on the impact velocity. Non-spherical agglomerates are also considered and, finally, the following question is addressed – 'How do agglomerates fracture?'.

Due to the short duration of an impact event, information from physical experiments is normally restricted to post-impact examinations of the fragments and debris produced. Explanations tend to rely on inferences that are based on solid mechanics concepts of brittle or semi-brittle fracture. However, it is not clear to what extent such solid mechanics ideas are applicable to particle systems such as agglomerates. Numerical simulations of systems of discrete particles are not restricted by small time or length scales and the micro-examination of short duration events such as fragmentation is ideally suited to DEM modelling.

The storage, transport, handling and processing of particulate materials constitutes a significant part of the operations in most chemical, pharmaceutical and allied industries. These particulate materials are frequently in the form of powders which are, themselves, agglomerations of much smaller sized primary particles. A common problem inherent in the handling of powders is the degradation resulting from attrition and/or fragmentation of the agglomerates as they collide with each other and with the process equipment.

Due to the short duration of an impact event, information from physical experiments is normally restricted to post-impact examinations of the fragments and debris produced. Explanations tend to rely on inferences that are based on solid mechanics concepts of brittle or semi-brittle fracture. However, it is not clear to what extent such solid mechanics ideas are applicable to particle systems such as agglomerates. Numerical simulations of systems of discrete particles are not restricted by small time or length scales and the micro-examination of short duration events such as fragmentation is ideally suited to DEM modelling.

C. Thornton, *Granular Dynamics, Contact Mechanics and Particle System Simulations*, Particle Technology Series 24, DOI 10.1007/978-3-319-18711-2_6

In all the simulations reported in this chapter the normal particle-particle interactions were modelled by JKR theory, see Sect. 3.3.1, and the tangential interaction model has been described in Sect. 3.3.2.

6.1 Agglomerate-Wall Impacts

6.1.1 2D Simulations

The first simulations of agglomerate impact fracture were performed by Yin (1992) and reported by Thornton et al. (1996). Due to computing limitations at the time, a 2D monodisperse, circular agglomerate consisting of 1000 primary particles was used. The particles were attributed with surface energy and the particle interactions were modelled by the adhesive-elastic contact force model described in Sect. 3.3.

The primary particles were initially randomly generated as a granular gas within a prescribed circular region and a centripetal gravity field was then imposed to bring the particles together. When the particles were satisfactorily packed together, surface energy was introduced and increased incrementally to the desired value. The centripetal gravity field was then slowly reduced and readjusted to provide a vertical gravity field with $g = -9.81$ m/s^2. The preparation stage was somewhat tedious and required continuous monitoring in order to ensure a satisfactory dense agglomerate. All particle velocities were then zeroed and the agglomerate was allowed to equilibrate. A wall was then created in a suitable location and then, by specifying a wall velocity for an appropriate number of timesteps, the wall was brought into a position within one timestep of contact with the agglomerate and its velocity was set to zero.

The radius and density of the primary particles was 100 μm and 2650 kg/m^3 respectively. The elastic properties of both the primary particles and the wall were $E = 70$ GPa and $\nu = 0.3$. The coefficient of interface friction was set at 0.35 and the surface energy was 3.0 J/m^2. The initial state of the agglomerate before impact is shown in Fig. 6.1.

The initial particle configuration, Fig. 6.1a, would appear to have a regular structure. The actual microstructure is more clearly illustrated by the equivalent space lattice, Fig. 6.1b, which is formed by connecting the centres of particles in contact by solid lines. The figure confirms the overall crystalline structure and highlights the structural defects. The space lattice is replotted to show the contacts subjected to compressive forces in Fig. 6.1c and the contacts at which the force is tensile in Fig. 6.1d. The compressive forces are of similar magnitude as the tensile forces and it is clear that both distributions are random.

To simulate the normal impact of the agglomerate with the wall, the desired impact velocity was then specified and all primary particles were attributed this initial value. At relatively high impact velocities, the interaction with the wall results in a compressive wave which propagates from the point of contact through the agglomerate. Behind the wavefront contacts are broken, contact sliding occurs and all but a few contacts are in compression. From an examination of the locations of sliding contacts and contacts that have been broken, the agglomerate appears to

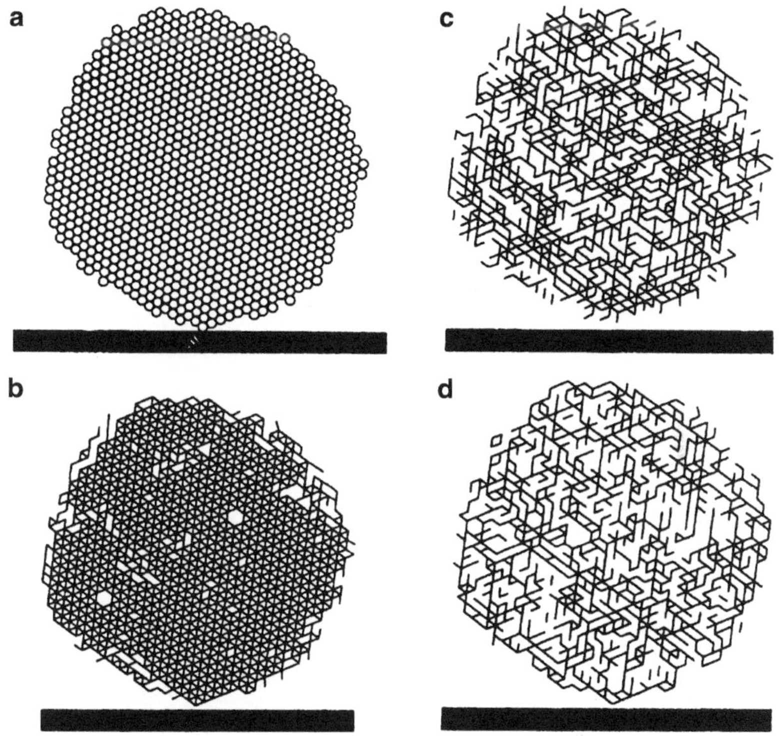

Fig. 6.1 Monodisperse 2D agglomerate – initial state (**a**) particle configuration (**b**) equivalent space lattice (**c**) compressive forces (**d**) tensile forces

deform plastically rather than exhibiting fracture. This plastic deformation process is illustrated in Fig. 6.2.

For an impact velocity of 1.0 m/s, Fig. 6.2 shows the space lattice after 6.5 μs when the wall force was still increasing. Figures 6.2a, b show the distributions of contacts carrying compressive and tensile forces respectively, which can be compared with Figs. 6.1c, d to reveal the changes that have occurred. As described above, behind the wavefront most of the contacts are in compression and the relatively few contacts carrying tensile forces are orientated in one of two directions. Figure 6.2 also shows the distributions of all surviving contacts at that time and the ones that have been broken.

The plastic deformation zone spreads throughout the agglomerate at high impact velocities and this results in a depleted structure with a coordination number of four. Then, as the tale-end of the compression wave travels through the agglomerate, more contacts are broken due to tensile separation. This results in shattering of the previously created plastic deformation zone. Once the contact force with the wall has reduced to zero the maximum number of contacts has been broken and the initial damage caused by the impact is complete.

If the impact velocity is reduced by an order of magnitude to 0.1 m/s then the behaviour is very different from that described above. Initially, a compression wave

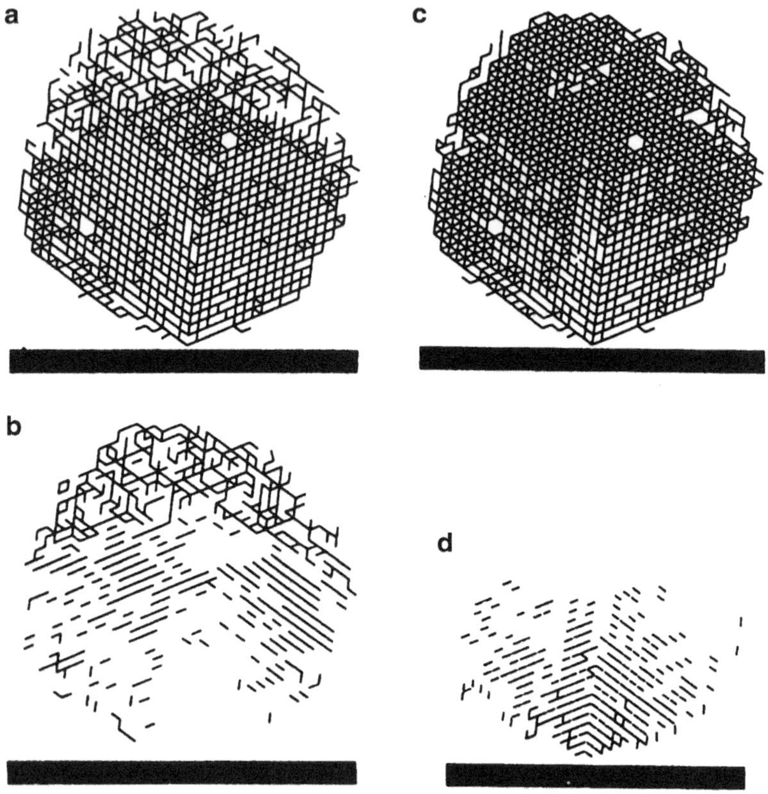

Fig. 6.2 Space lattice after 6.5 μs (max. wall force occurred at 10 μs) for V = 1.0 m/s (**a**) compressive forces (**b**) tensile forces (**c**) existing contacts (**d**) deleted contacts

propagates from the wall producing sliding and contact breaking associated with plastic deformation. However, since the force generated at the wall is much smaller the effect is less extensive. Due to attenuation of the stress wave, the plastic deformation zone is restricted to a small localised region near the point of impact. This is followed by the initiation of cracks at the perimeter of the plastic deformation zone which propagate outwards towards the agglomerate surface. By the time that the force on the wall has reduced to zero, the initial damage is complete and the crack pattern is established. There is then a relatively long period of crack opening inwards from the agglomerate surface.

At sufficiently low impact velocities no crack formation occurs and the agglomerate rebounds. Damage, however, may be produced as a result of localised plastic deformation adjacent to the wall contact. Also, due to the centre of mass of the agglomerate not being exactly vertically above the point of impact with the wall, a rotation was imparted to the agglomerate as it rebounded from the wall. Therefore, at sufficiently low impact velocities, the agglomerate behaved similar to that of a single, almost elastic, sphere or disc.

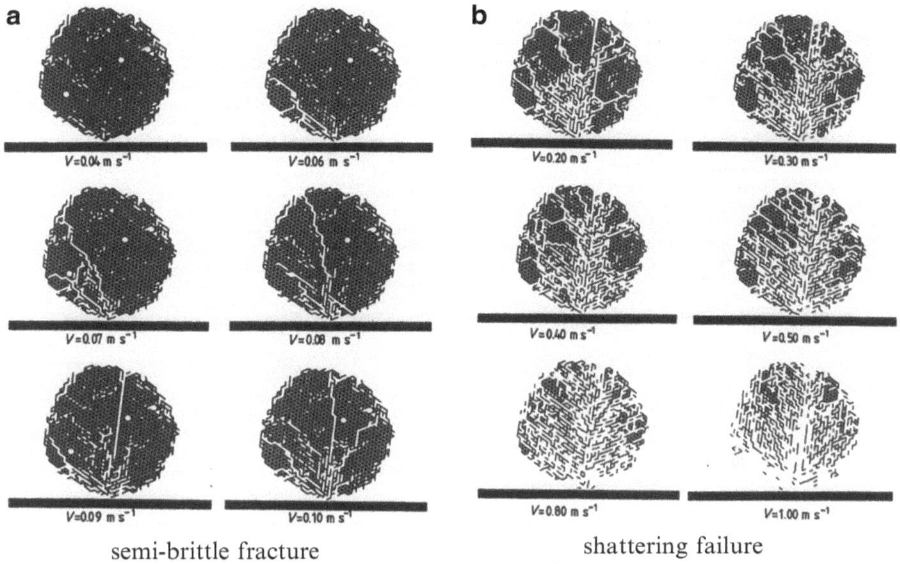

Fig. 6.3 Damage patterns

The agglomerate damage pattern (involving plastic deformation, shattering or crack initiation and propagation) is completed at the time when the wall force becomes zero. This is followed by a long period during which the established cracks open and would be visible to the conventional observer. The initial damage pattern defines the damage caused by the impact with the wall and is best illustrated by the equivalent space lattice. Figure 6.3 shows the equivalent space lattice for different impact velocities after an elapsed time of 100 μs. For an impact velocity of 1.0 m/s, most of the contacts are broken as a result of the extensive plastic deformation zone and the agglomerate, as a consequence, is shattered. The extent of the plastic deformation zone reduces with decreasing impact velocity. At an impact velocity of 0.1 m/s, the plastic deformation zone is contained to a small region near to the point of impact. From the perimeter of the plastic zone, cracks propagate to the surface of the agglomerate leading to a semi-brittle fracture mechanism. It can be seen in the figure that, for semi-brittle behaviour, the extent of fracture damage is very sensitive to small changes in impact velocity. At an impact velocity of 0.04 m/s, no cracks are observed but a small amount of internal damage has occurred due to localised plastic deformation.

In Fig. 6.3 the elapsed time when the wall force reduced to zero was 53 μs for an impact velocity of 0.1 m/s and 17 μs when the impact velocity was 1.0 m/s. For both cases the particle configuration is shown in Fig. 6.4 at a time of 1.3 ms. The figure clearly illustrates the difference between semi-brittle fracture and failure due to shattering.

Of practical interest is the size distribution of the daughter fragments resulting from a collision. It is conventional to show the cumulative percentage mass of

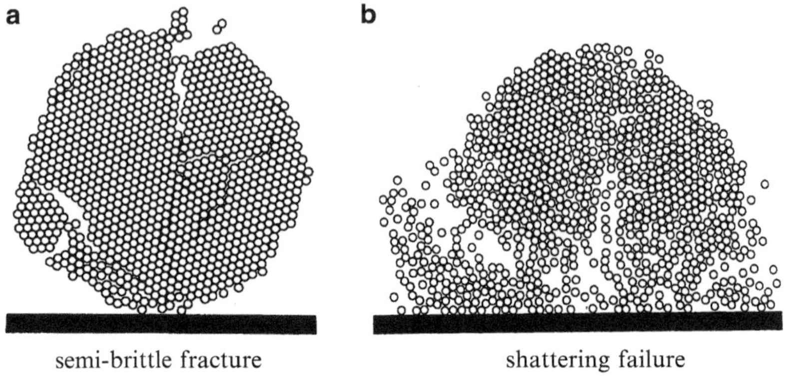

Fig. 6.4 Particle configurations for (**a**) V = 0.1 m/s (**b**) V = 1.0 m/s

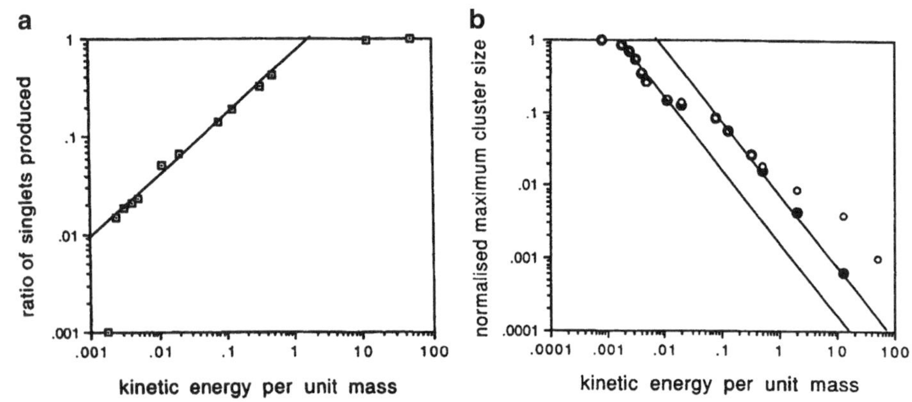

Fig. 6.5 Effect of specific energy input on (**a**) the number of singlets produced and (**b**) the size of the largest surviving fragment

material undersize plotted against size. However, in comparison with the 3D situation, large 2D arrays are 'small' in terms of their mass. As a consequence, the particle size distribution curves obtained from the simulations, when represented on a double-logarithmic plot, do not exhibit the bilinear characteristics which distinguish between the 'residue' of the larger fragments and the 'complement' made up of the smaller debris, as found experimentally by Arbiter et al. (1969) for impact velocities that produced semi-brittle fracture. We may, however, examine the extremes of the distribution, namely the percentage of singlets produced and the size of the largest fragment.

If we normalise the number of singlets produced, by dividing by the total number of constituent particles in the agglomerate prior to impact, and define the specific energy input as the initial kinetic energy per unit mass then, as shown in Fig. 6.5a, it is found that the percentage of singlets is a power law function of the specific energy with an exponent of 5/8. Deviations from this relationship are only observed at very high velocities that produce more than 90 % of singlets and for an impact velocity of 0.06 m/s, which produced one singlet only. The simulation data satisfies the above relationship over a range of impact velocities that include both semi-brittle fracture and shattering.

In order to examine the maximum cluster size (largest fragment), two definitions are used. The first is simply to count the number of primary particles in a fragmentation cluster and divide by the number of primary particles in the original agglomerate. Such a cluster may include a string of particles that remains attached to the main body of the cluster by a single bond and, in terms of survival, this definition is optimistic. Data obtained using this definition is represented in Fig. 6.5b by open circles.

An alternative is to consider the rigidity of such clusters. If three spheres are connected by bonds to form a triangular arrangement then they form a rigid structure that cannot be deformed, ignoring any elastic deformation, without breaking a bond. Therefore an alternative definition is to define a rigid cluster by the number of continuously connected triangles formed by the particles as represented by the equivalent space lattice. Only triangles connected by a common side are accepted. If two triangles are connected at their common apex then they are considered to belong to different clusters. The size of the cluster so defined is normalised by dividing by the number of the underlying rigid cluster of the original agglomerate. In Fig. 6.5b the solid circles represent data obtained using this alternative definition of cluster size.

Figure 6.5b illustrates the relationship between the normalised maximum cluster size and the specific energy. This representation has been used in asteroid physics by Takagi et al. (1984) who examined the impact fragmentation of centimetre-sized rock particles and obtained a power law relationship similar to that in Fig. 6.5. It can be seen from the figure that the results are not significantly affected by the size definition except at very high velocities when the maximum cluster consists of less than ten primary particles, which are not necessarily compact. Therefore, only the results based on the size of rigid clusters, i.e. the solid circles, will be considered further. The data is best described by two straight lines that indicate that the maximum cluster size is inversely proportional to the specific energy. As the maximum cluster size approaches 10 % of the size of the initial agglomerate there is a transition from one line to the other. This transition regime corresponds to a range of impact velocities between 0.15 and 0.40 m/s and an examination of Fig. 6.3 confirms that the transition between the semi-brittle and shattering regimes occurs over this narrow velocity range.

6.1.2 3D Simulations

In this subsection agglomerate breakage is illustrated by computer generated images of the configuration of the primary particles, in which the particles are colour coded to indicate the size of the fragment to which they belong. Fracture planes are more clearly identified by representing an agglomerate by the equivalent space lattice formed by connecting the centres of particles in contact by solid lines. However, in order to visualise the evolution of bond breakage leading to fracture, it is more appropriate to use solid lines to connect the centres of particles that were initially in contact but which have broken contact during the impact.

6.1.2.1 Impact of a Crystalline Agglomerate

Kafui and Thornton (2000) reported the results of 3D simulations of a crystalline (face-centred cubic) agglomerate impacting normal to a horizontal elastic target wall. The agglomerate consisted of 7912 spheres of diameter $d_p = 20$ μm and had an overall diameter $d_a = 0.461$ mm with an average coordination number $Z = 11.1$ after preparation; $Z < 12$ due to the finite size of the agglomerate. The elastic properties of both the primary particles and the wall were $E = 70$ GPa and $\nu = 0.35$. The solid density of the primary particles $\rho = 2650$ kg/m^2 and the interface friction coefficient $\mu = 0.3$. Five values of interface energy $\Gamma = 0.2$, 0.4, 1.0, 2.0 and 4.0 J/m^2 were used and impact was carried out at velocities in the range 0.05–20 m/s.

Having a face-centred cubic microstructure, impact fracture depends on the microstructural orientation of the packing. For the results shown in this subsection the orientation of the face-centred cubic packing was such that the square-packed planes were vertical.

The evolution of bond breakage during the impact is illustrated in Fig. 6.6 by the space lattice showing only the broken bonds. Two orthogonal views (an elevation and a view from above) are shown for each elapsed time. During loading, bonds are broken along four vertical planes that form a diamond shaped pattern when viewed from above. The four planes coincide with the perimeter of the agglomerate-wall interface and are the consequence of relative shear motion between adjacent load-transmitting and load-free dense (triangular) packed vertical planes. The relative shear motion results in the breaking of one set of contacts between the loaded and unloaded dense-packed planes. The bond breaking propagates upwards from the wall until it reaches the top of the agglomerate when the kinetic energy attains its minimum value at $t = 1.065$ μs. During unloading, as the stored elastic energy is converted into kinetic energy, a second set of contacts are broken, propagating downwards from the top of the agglomerate. With further increase in kinetic energy, some bonds are also broken in the third set of contacts leading to fracture along some of the shear-induced weakened planes just prior to the end of the impact. For an impact velocity of 2.0 m/s with interface energy of 4.0 J/m^2, Fig. 6.7 shows the space lattice after the end of the impact.

The effect of impact velocity on the breakage pattern obtained is shown in Fig. 6.8 for agglomerates with (a) $\Gamma = 4.0$ J/m^2 and (b) $\Gamma = 0.4$ J/m^2. Figure 6.8a

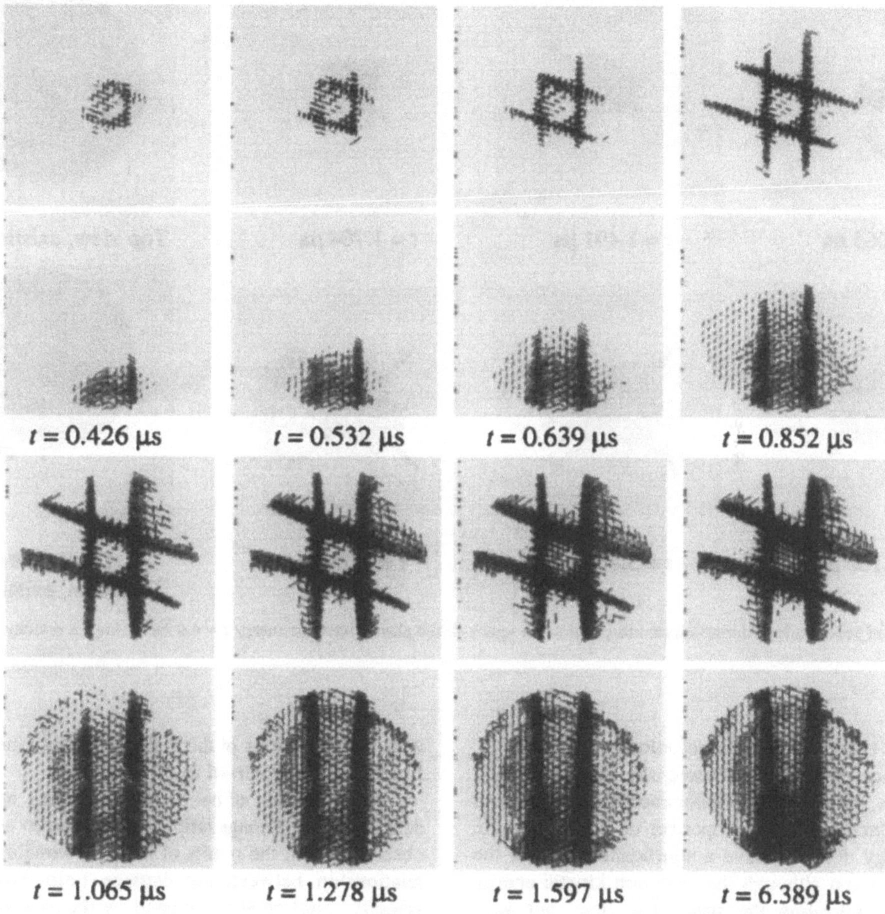

Fig. 6.6 Evolution of bond breakage for an impact velocity $V = 2.0$ m/s and interface energy $\Gamma = 4.0$ J/m^2

clearly shows that at an impact velocity of 5.0 m/s all four of the shear-induced weakened planes are fractured plus two short fracture planes that are parallel to the square–packed planes. At lower impact velocities the fracture pattern is a subset of that obtained for $V = 5.0$ m/s. If the impact velocity is increased above 5.0 m/s no extra fracture planes are created but the residual fragments are weakened due to internal bond breakage and this leads to shattering at high velocities. For the weak agglomerate shown in Fig. 6.8b, varying the impact velocity produced similar results but at much lower impact velocities. Figure 6.9 shows that similar fracture patterns are obtained for $\Gamma = 2.0$ J/m^2. Also shown in the figure are images of the corresponding particle configurations, as viewed from below, which illustrate the

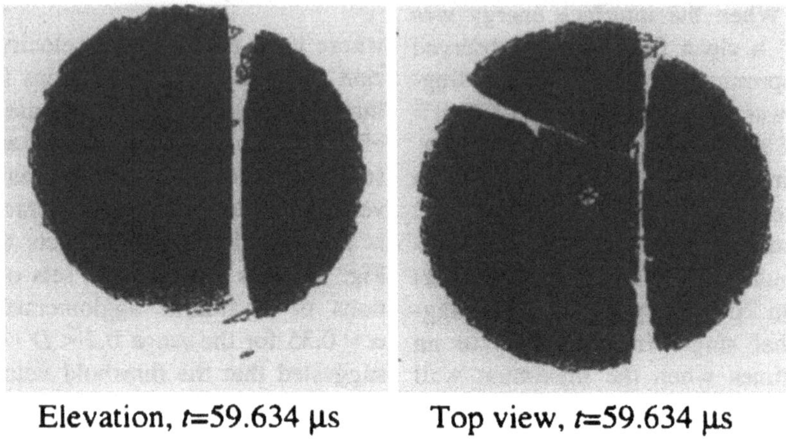

Fig. 6.7 Space lattice after the end of impact

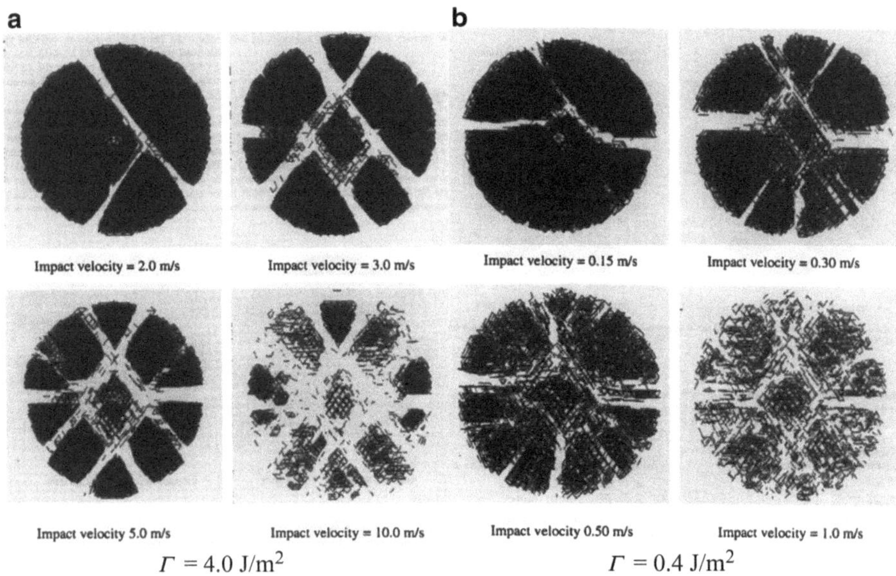

Fig. 6.8 Equivalent space lattices viewed from below illustrating the breakage patterns for (**a**) a strong and (**b**) a weak agglomerate

increasing amount of fine debris produced around the impact area when the impact velocity is increased. For a velocity of 3.0 m/s, the largest surviving fragment consisted of only 312 primary particles.

To quantify the internal damage, the proportion of bonds broken during an impact can be defined as the damage ratio D, which is plotted against impact

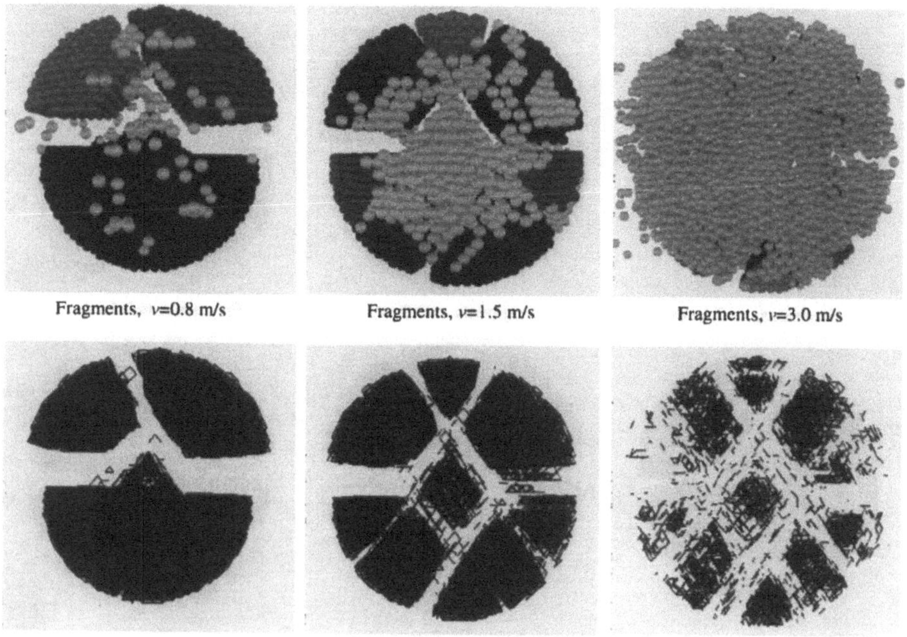

Fig. 6.9 Configuration of primary particles (*top*) and equivalent space lattice (*bottom*) as viewed from below, for a strong agglomerate ($\Gamma = 2.0$ J/m^2)

velocity in Fig. 6.10 for different values of interface energy. All the data sets can be approximated by the straight lines defined by

$$D = \alpha \ln\left(\frac{V}{V_0}\right) \tag{6.1}$$

where V_0 is the threshold velocity below which no significant damage occurs. Deviations from Eq. (6.1) occur at the limits $D \rightarrow 0$ and $D \rightarrow 1$ because (a) there will be a range of low velocities at which the agglomerate rebounds but suffers a small amount of internal damage and (b) even at very high velocities not all contacts will be broken since it is possible that some doublets and triplets will survive.

The data sets shown in Fig. 6.10 satisfy Eq. (6.1) with $\alpha = 0.35$ for the range $0.2 < D < 0.8$. The threshold velocity increases with interface energy, as shown in Fig. 6.11, which can be expressed by the following power law.

$$V_0 = 0.17\Gamma^{3/2} \tag{6.2}$$

This implies that the damage ratio should scale with $\ln(V/\Gamma^{3/2})$ and this is confirmed reasonably well in Fig. 6.12 except for the results of the weakest agglomerate.

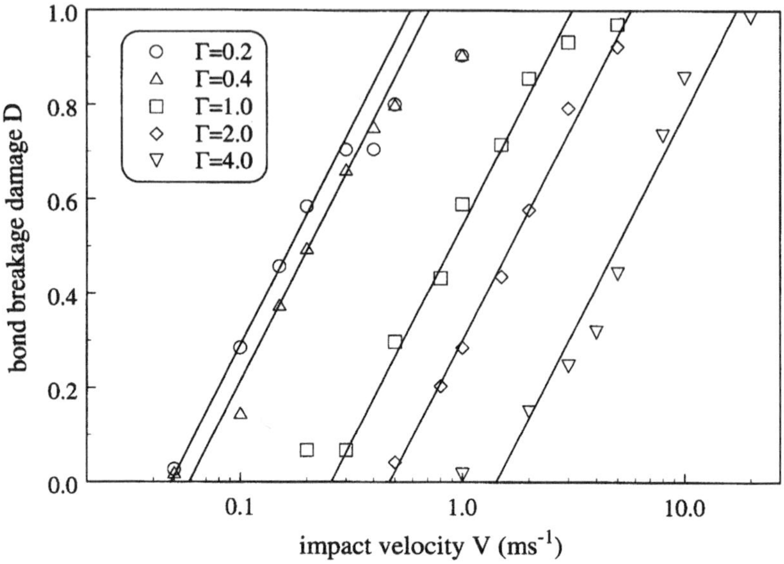

Fig. 6.10 Effect of impact velocity on the damage ratio D

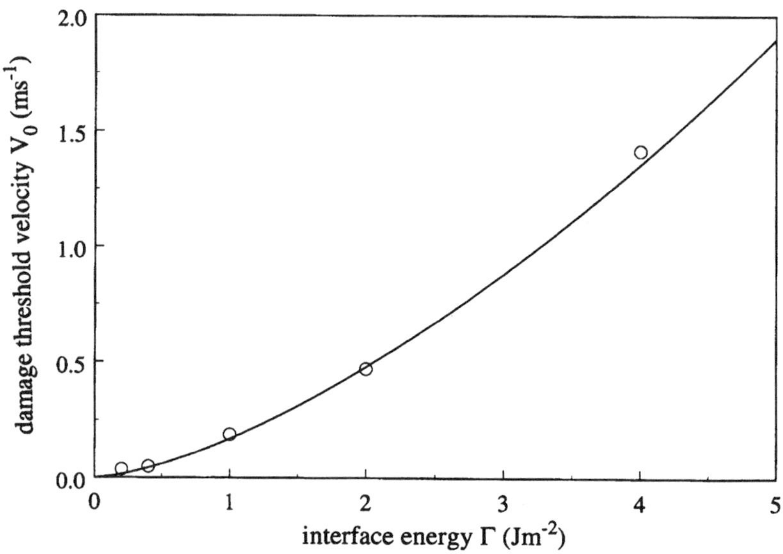

Fig. 6.11 Relationship between threshold velocity V_0 and interface energy

The size distributions of the fragments produced by impact breakage are illustrated in Fig. 6.13 for (a) a weak agglomerate and (b) a strong agglomerate. Using a double-logarithmic plot of the cumulative mass fraction undersize f against normalised size m/M where m is the mass of a cluster and M is the mass of the original agglomerate. The figure shows that the fragment size distributions exhibit

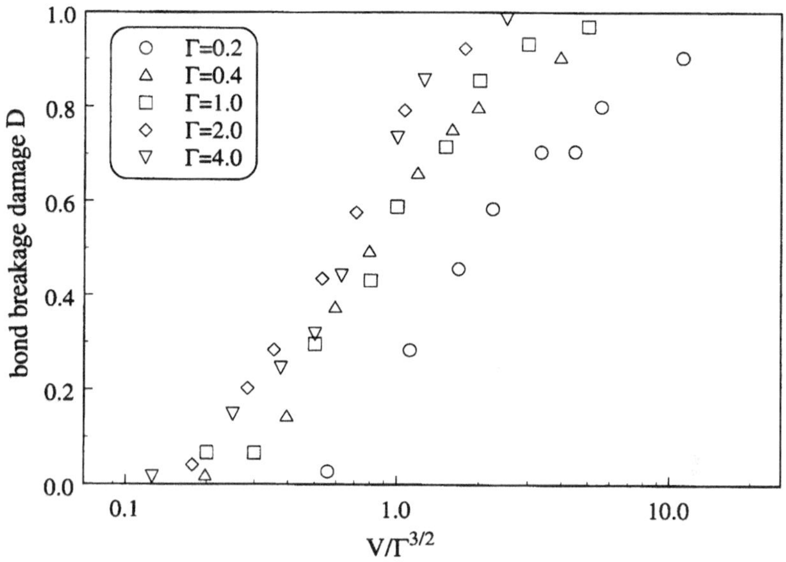

Fig. 6.12 Relationship between damage ratio, impact velocity and interface energy

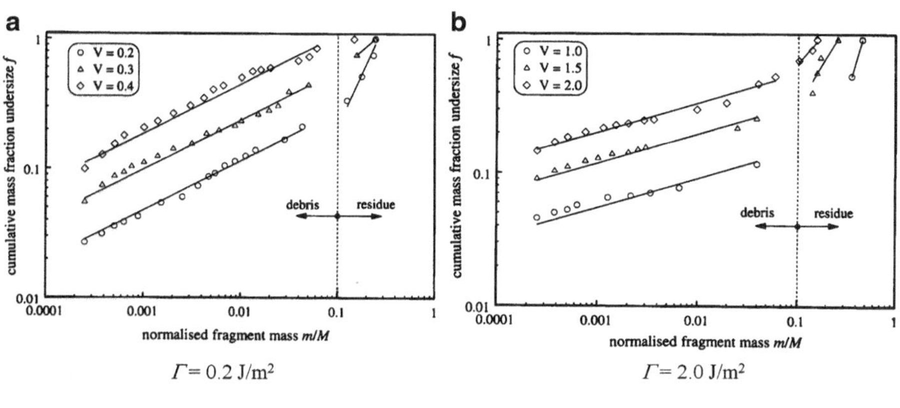

Fig. 6.13 Effect of impact velocity on the fragment size distribution of (**a**) weak and (**b**) strong agglomerates

bilinear characteristics, which distinguish the residue of large fragments from the complement of small fragments (debris). Similar results were obtained from free-fall impact tests on sand-cement and limestone-cement spheres by Arbiter et al. (1969) for impact velocities that produced semi-brittle fracture.

Considering the debris (complement), Fig. 6.13 shows that the mass fraction undersize increases with impact velocity. Arbiter et al. (1969) demonstrated that the

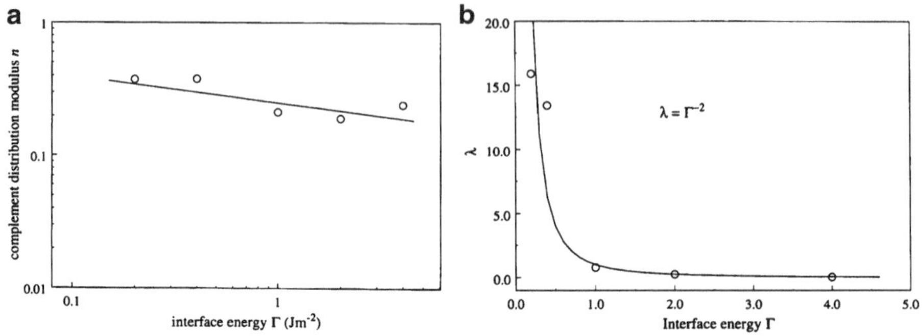

exponent n prefactor λ

Fig. 6.14 Effect of interface energy on the complement size distribution parameters

size distribution of the debris (complement) correlates with the specific impact energy according to the expression

$$f = \lambda V^2 \left(\frac{m}{M}\right)^n \tag{6.3}$$

As indicated in Fig. 6.13, both the exponent n and the prefactor λ depend on the interface energy Γ. The two relationships are provided in Fig. 6.14 and may be approximated by

$$n = 0.25\Gamma^{-1/5} \tag{6.4}$$

and

$$\lambda = \Gamma^{-2} \tag{6.5}$$

Using Eqs. (6.4) and (6.5), the normalised fragment size distribution of the debris is plotted in Fig. 6.15. It can be seen that all the data collapse reasonably well onto a single curve. In general, however, λ is expected to depend also on other properties such as solid fraction and coordination number which define the microstructure.

The size of the largest surviving fragment m_L is plotted against the ratio of impact velocity to interface energy in Fig. 6.16 which shows that the data satisfies a power law scaling given by

$$\frac{m_L}{M} = 0.15\left(\frac{V}{\Gamma}\right)^{-3/2} \tag{6.6}$$

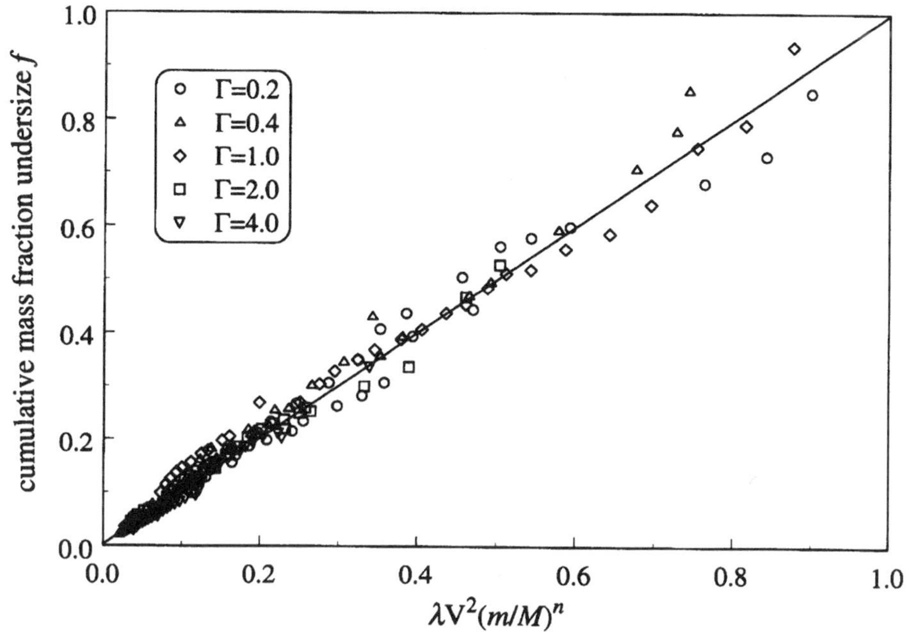

Fig. 6.15 Normalised fragment size distribution of the debris (complement)

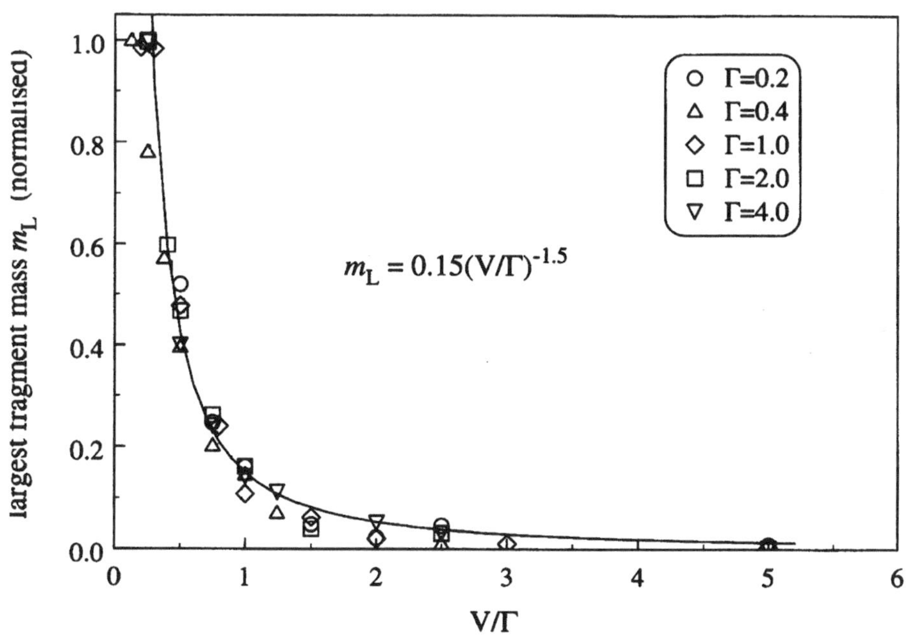

Fig. 6.16 Normalised mass of the largest fragment

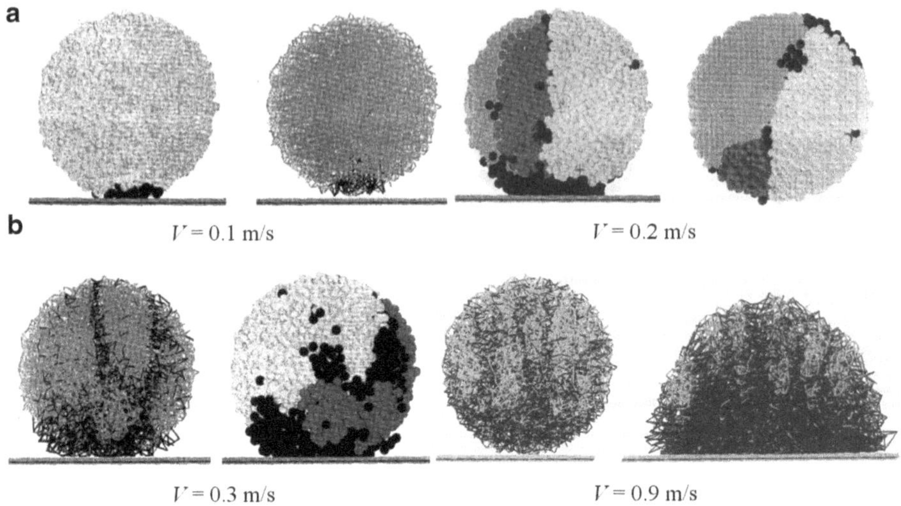

Fig. 6.17 Examples of agglomerate damage

6.1.2.2 Impact of Polydisperse Agglomerates

To prepare polydisperse agglomerates, the primary particles are randomly generated as a granular gas within a defined region, spherical, cuboidal, cylindrical or any other desired shape. A centripetal gravity field is then imposed to bring the particles into contact. During this stage the interparticle friction is set to zero. The duration of the imposed centripetal gravity field determines the solid fraction within the agglomerate. When the desired solid fraction is reached, each particle is assigned with the desired values of surface energy and interparticle friction. Having formed an agglomerate with the requisite properties, the centripetal gravity field is slowly reduced to zero to complete the preparation stage.

Thornton et al. (1999) reported simulations of a dense polydisperse agglomerate consisting of 4000 primary particles of sizes in the range 60 ± 3 μm. The average diameter of the agglomerate was 1.113 mm with a solid fraction of 0.653 and an average coordination number of 4.879. Examples of the damage produced at the end of the impacts for different impact velocities are shown in Fig. 6.17.

For an impact velocity of 0.1 m/s, the agglomerate rebounded from the target wall and did not fracture. Only a small percentage of bonds were broken, these were concentrated close to the impact zone and resulted in a small amount of fine debris. The total number of debris particles was 36, which corresponds to slightly less than 1 % of the initial agglomerate mass, and the total number of broken bonds was 223 compared to the initial 9,758 bonds. When impacted at a velocity of 0.2 m/s, the agglomerate fractured into three (two large and one medium-sized fragments). At a velocity of 0.3 m/s, there were six surviving fragments at the end of the test and much more bond breakage and debris produced. The surviving fragments were not just smaller but also weaker, due to more internal damage. The agglomerate

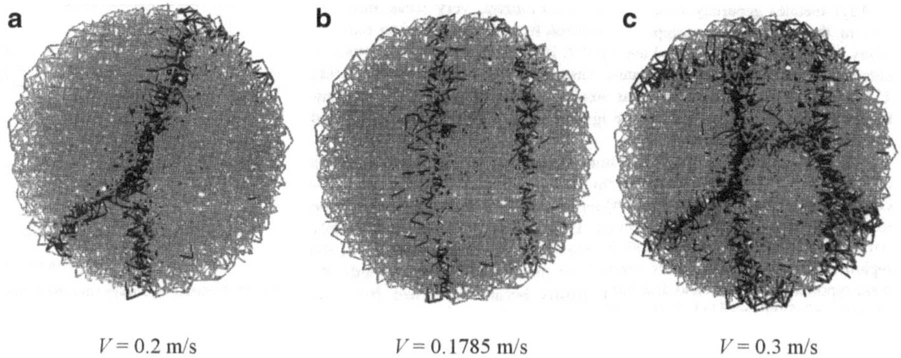

Fig. 6.18 Space lattice views from above at the end of impact, showing existing contacts (*grey*) and broken contacts (*black*)

shattered when impacted at a velocity of 0.9 m/s. The high impact velocity at first tended to break-up the agglomerate into many medium-sized fragments as illustrated by the space lattice after 9 μs when the platen force had started to reduce. The space lattice at the end of the test ($t = 512$ μs) is also shown, indicating the extensive bond breakage and flattening of the agglomerate. The largest fragment at the end of the test contained only 153 particles.

Figure 6.18 provides a comparison of fracture patterns obtained for three impact velocities. It is apparent, from Figs. 6.18a, b that a small variation in impact velocity resulted in significantly different fracture patterns. However, it would appear from the figure that the two fracture patterns are essentially subsets of the fracture pattern created by the higher impact velocity of 0.3 m/s, as shown in Fig. 6.18c. In the case of crystalline agglomerates it was clear that during the loading stage there is a shear induced weakening of certain planes and then, during unloading, a number of these pre-conditioned planes fractured and the extent of the fracture pattern depended on the magnitude of the impact velocity. The same might be expected in the case of polydisperse agglomerates but it is more difficult to verify this hypothesis.

From agglomerate impact simulations, Mishra and Thornton (2001) found that loose agglomerates never fractured. It is therefore necessary to redefine the terminology used to describe the observed breakage phenomena. The term "fracture" is reserved for breakage patterns in which clear fracture planes (cracks) are visible. This mode produces two or more large daughter fragments and is normally accompanied by some fines production adjacent to the impact site. If for example, due to the high impact velocity used, the large daughter fragments are themselves broken into small clusters of primary particles then the term "shattering" is used. An alternative mode of breakage is one in which there is no evidence of any attempted fracture and the end products consist of one cluster centred in the upper part of the agglomerate with the remainder of the agglomerate reduced to very small clusters of primary particles and singlets. This type of breakage is termed "disintegration". If the impact velocity is sufficiently high that disintegration extends throughout the agglomerate and there is no 'large' surviving cluster then this mode is referred to as "total disintegration". In this case, the size distribution of the fragments may be similar to that produced by shattering but the distinction is the difference in kinetic energy of the system at the end of the impact. When shattering occurs a significant number of small daughter fragments are projected at relatively high speeds away from the impact location. On the other hand, if total disintegration occurs the agglomerate simply collapses into a heap on the target wall.

Mishra and Thornton (2001) found that for 'compact' agglomerates (as opposed to fractal agglomerates) dense agglomerates always fracture and loose agglomerates always disintegrate. This was found to be true irrespective of the strength of the bonds between the primary particles which simply dictated the range of velocities over which breakage would occur. It was also found that either fracture or disintegration, or both, could occur for agglomerates with intermediate packing densities. It was demonstrated that, in the intermediate case, the mode of failure could change from disintegration to fracture by changing the location on the agglomerate surface that is used as the impact site.

Moreno et al. (2003) reported DEM simulations of oblique impacts of spherical agglomerates. They found that, for a constant impact speed, the number of bonds broken and the amount of debris produced decreased as the impact angle became more oblique. They demonstrated that this was due to the decrease in the normal velocity component and concluded that "the normal component of the impact speed is the dominant factor controlling the breakage of contacts". However, it was also shown that, for the same number of bonds broken, the spatial distribution of damage (broken bonds) depended on the impact angle.

For non-spherical agglomerates, breakage depends also on the orientation of the agglomerate prior to impact. To illustrate this, three agglomerates were prepared – one spherical, one cuboidal and one cylindrical. Each agglomerate consisted of 10,000 primary particles of sizes in the range 20 ± 4 μm. The final, as prepared, porosities of the spherical, cuboidal and cylindrical agglomerates were 0.395, 0.412 and 0.400 respectively, with corresponding coordination numbers of 5.124, 4.558 and 4.562, corresponding to 25,521, 21,831 and 22,092 contacts. The dimensions of the three agglomerates were 0.54 mm diameter (spherical), 0.480 mm $\times$ 0.477 mm $\times$ 0.484 mm (cuboidal) and 0.500 mm diameter $\times$ 0.474 mm length (cylindrical).

All the impacts are collinear normal impacts. That is to say that the line orthogonal to the target wall at the initial point of contact passes through the centre of mass of the agglomerate. In all cases an impact velocity of 1.0 m/s was used and the interface energy was specified as $\Gamma = 1.0$ J/m^2.

The spherical agglomerate was used as a benchmark against which the other agglomerates could be compared. The breakage of the spherical agglomerate is illustrated in Fig. 6.19, which shows that the agglomerate fractured; resulting in three large fragments (consisting of 4,990, 2,256 and 1,084 primary particles) plus a significant amount of small debris adjacent to the impact site.

Three different impact sites were selected for the cuboidal agglomerate – a face, an edge and a corner of the agglomerate. For the face impact, illustrated in

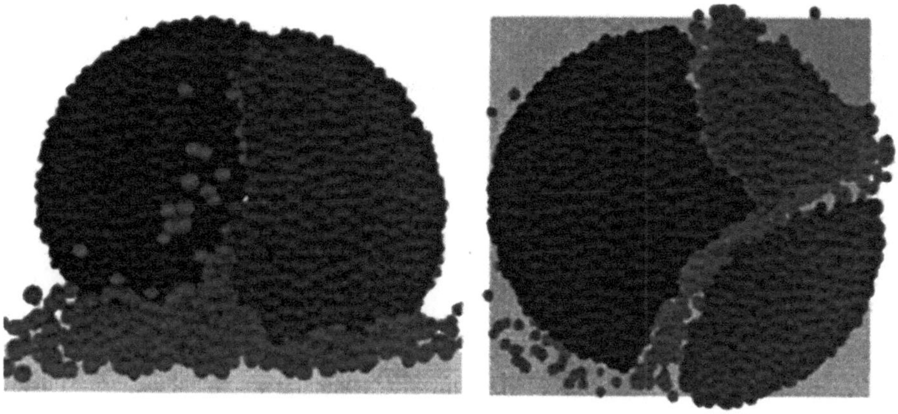

Fig. 6.19 Fragments for impact of the spherical agglomerate

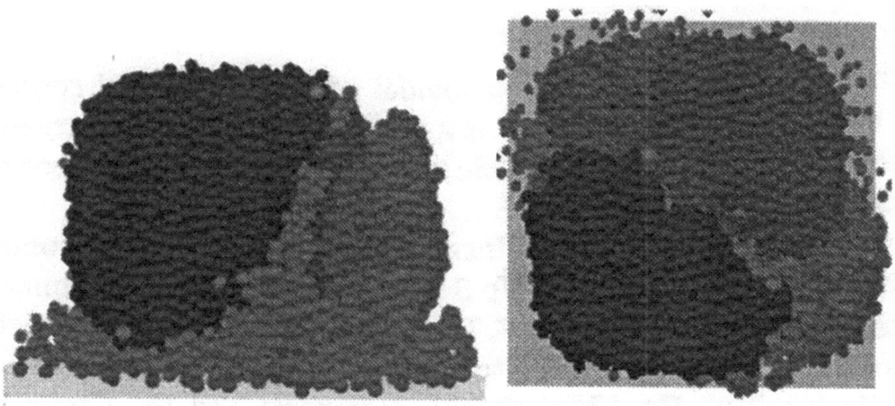

Fig. 6.20 Fragments for face impact of the cuboidal agglomerate

Fig. 6.20, the agglomerate fractured into four large fragments (3,045, 2,843, 1,114 and 801 primary particles) together with small debris due to disintegration adjacent to the wall. However, as can be seen in Fig. 6.21a, fracture did not occur when the agglomerate impacted the wall along one of its edges. Disintegration adjacent to the wall produced small debris with a largest cluster of 15 primary particles. The large surviving cluster consisted of 9,030 primary particles with no evidence of any internal damage. A similar breakage pattern occurred for the corner impact, see Fig. 6.21b, but with a lower degree of disintegration adjacent to the wall.

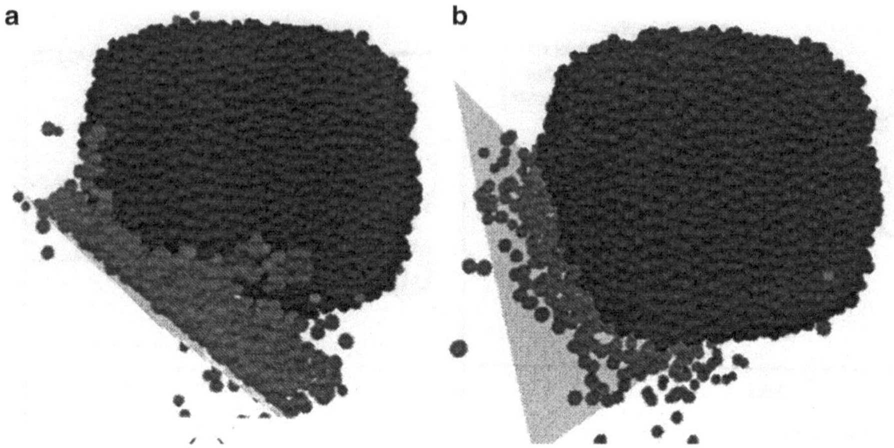

Fig. 6.21 Fragments for (**a**) edge and (**b**) corner impacts of the cuboidal agglomerate

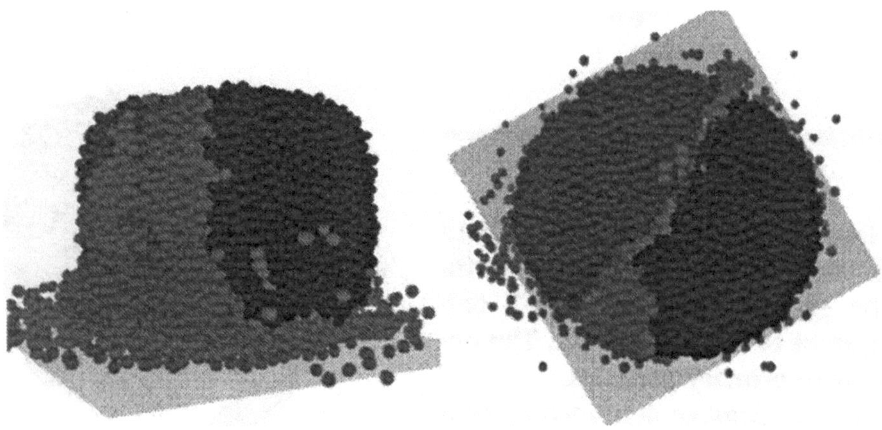

Fig. 6.22 Fragments for circular end impact of the cylindrical agglomerate

For the cylindrical agglomerate, impact sites were selected to provide a circular end impact, a side impact and a rim impact. The resulting breakage patterns are shown in Figs. 6.22 and 6.23.

It can be seen that both the end impact and the side impact resulted in fracture but fracture was not observed for the rim impact. In the case of the end impact, Fig. 6.22, the agglomerate fractured into two large fragments (3,975 and 3,186 primary particles) and a medium sized fragment (607 primary particles) resulting from the bifurcation of the primary fracture; the remaining damage being small debris adjacent to the wall. In the side impact, see Fig. 6.23a, aside from the small

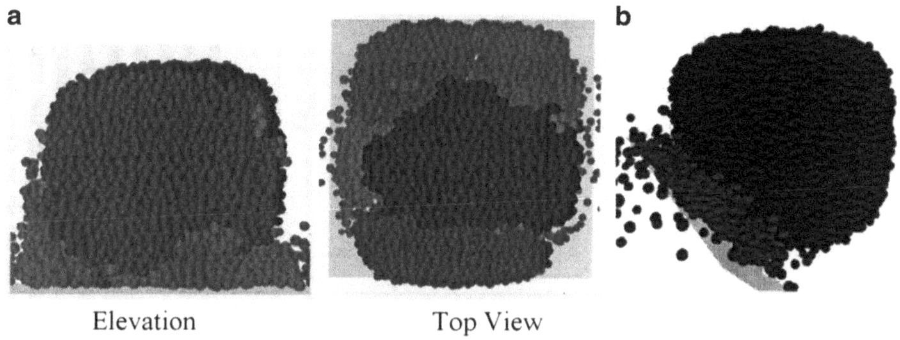

Fig. 6.23 Fragments for (**a**) side and (**b**) rim impacts of the cylindrical agglomerate

debris, fracture resulted in four large fragments (2,818, 2,012, 1,608 and 1,228 primary particles). Fracture did not occur when the impact was against the rim of the agglomerate, as shown in Fig. 6.23b. The one large surviving cluster consisted of 9,210 primary particles.

The above illustrations of agglomerate damage demonstrates that, for the agglomerate specification in terms of number of primary particles, bond strength and impact velocity, disintegration always occurs adjacent to the impact site and that this is where the small debris is produced. Whether fracture occurs or not depends on the impact site location. If fracture occurs then the fracture pattern and the consequent size and shape of the large surviving fragments depend on both the agglomerate shape and the location on the agglomerate surface used as the impact site.

For all the impacts, the time evolution of the damage ratio is shown in Fig. 6.24. It can be seen that the data sets for the non-spherical agglomerates fall into two groups. Group A consists of the cubical face, cylindrical end and cylindrical side impacts, all of which attain a final damage ratio of 0.32 which is significantly higher than the value of 0.205 obtained for the spherical agglomerate. In contrast, the final damage ratio for all Group B impacts is about 0.185 which is slightly less than that of the spherical agglomerate. The figure shows that the damage ratio increases rapidly to the final asymptotic value except for the cuboidal edge and cuboidal corner impacts. In these two cases, there is a delay before the rapid increase in damage ratio occurs, most notable in the case of the cuboidal corner. From an examination of the time evolution of the wall force and the number of wall contacts, see Liu et al. (2010), it was found that the number of wall contacts monotonically increased except for the cuboidal corner, cuboidal edge and cylindrical rim impacts. In these Group B impacts the number of wall contacts initially increased to a maximum value and then decreased to a constant asymptotic value. Although there were significant fluctuations in the evolution of the corresponding wall forces, the general trend for Group A impacts was that of an increase to a maximum force in about 10 μs followed by a reduction at a decreasing rate to a negligible value corresponding to the self-weight of the residual fragments. In contrast, for Group B

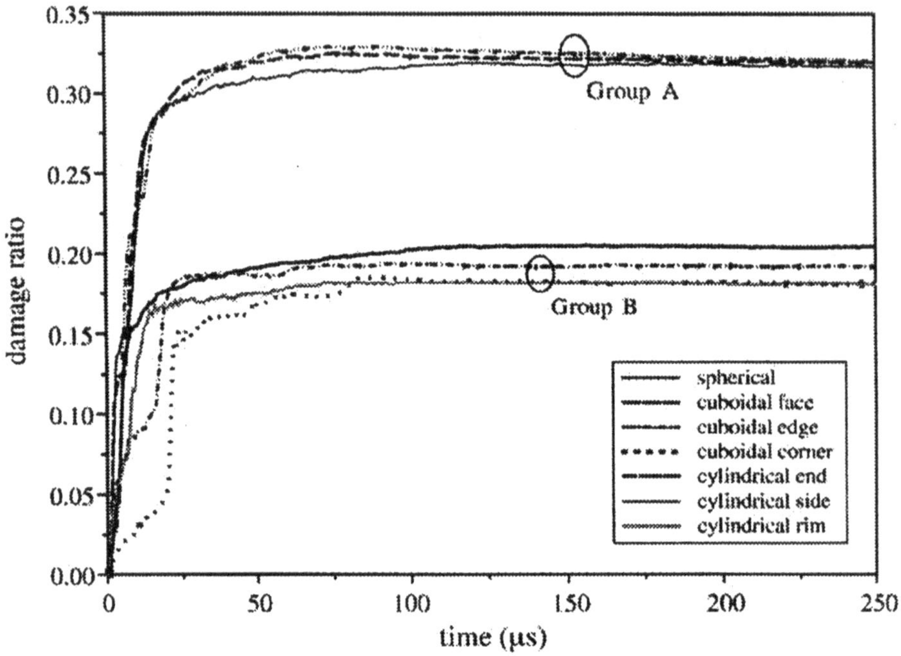

Fig. 6.24 Evolution of damage ratio during impact

impacts, there was a delay before the wall force increased to a maximum value. This was most notable for the cuboidal edge and corner impacts. In both cases, the peak wall force coincided with the rapid increase in broken contacts shown in Fig. 6.24.

The fragment size distributions resulting from all the impacts are shown in Fig. 6.25, which shows the expected bilinear distributions that distinguish the large fragments (residue) from the complement of small fragments (debris). The figure demonstrates that, for a given impact velocity, the amount of debris produced is dependent on agglomerate shape and impact site. The smallest amount of debris is produced by the cuboidal corner impact and the largest amount occurs for the cuboidal face impact. It is noted that the exponent of the debris is independent of agglomerate shape and impact site location, in this case about 0.13. It therefore follows that the exponent for the debris only depends on the bond strength, as illustrated in Fig. 6.13.

The cuboidal agglomerate was reoriented, as shown in Fig. 6.26a, so that it impacts the wall along the leading edge. The vector connecting the point of contact with the centre of mass of the agglomerate is inclined at 30° to the vertical. An initial vertical velocity of 1.0 m/s was specified for all the primary particles in order to simulate a non-collinear normal impact with the wall, which resulted in breakage of the agglomerate as shown in Fig. 6.26b. Figure 6.26b shows a thin central section (approximately three particles wide) in order to clearly illustrate the fracture mode.

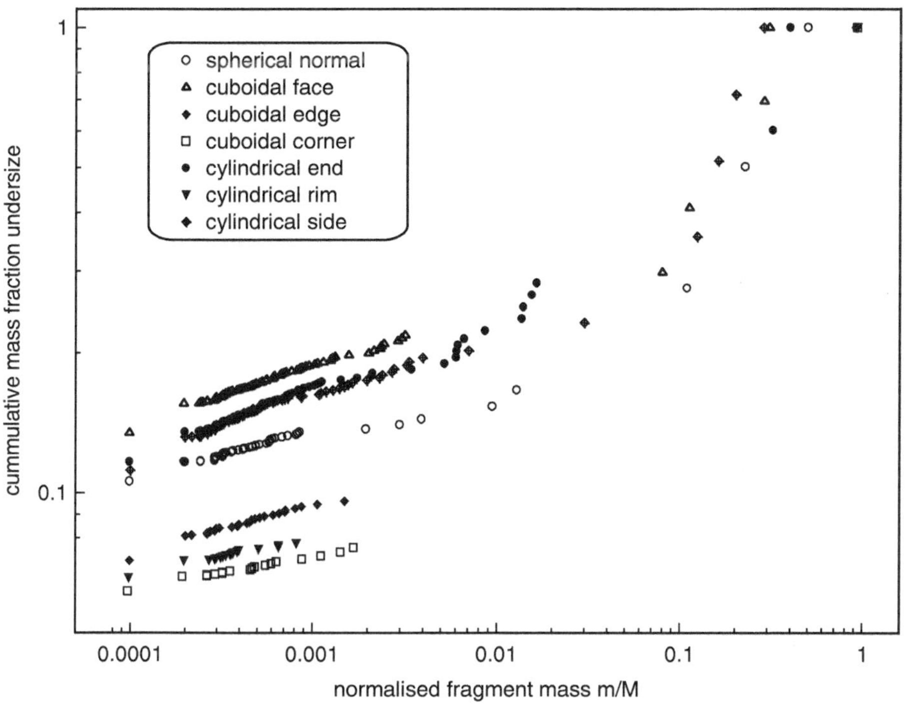

Fig. 6.25 Fragment size distributions

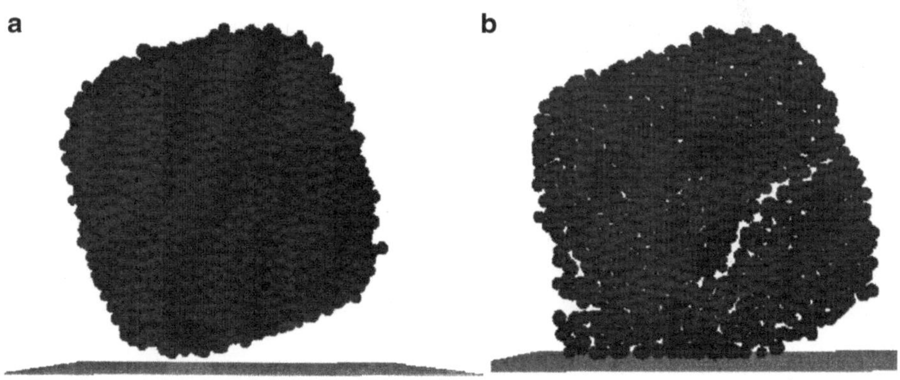

Fig. 6.26 Cuboidal agglomerate (**a**) before impact (**b**) after 11 μs

The advantage of this configuration is that, although the particle arrangement is three-dimensional, the overall behaviour is essentially two-dimensional making visualisations of the mechanisms much easier to identify.

The evolution of the total wall force generated by the impact is shown in Fig. 6.27. The wall force increases to 6.5 mN, drops and then increases to a

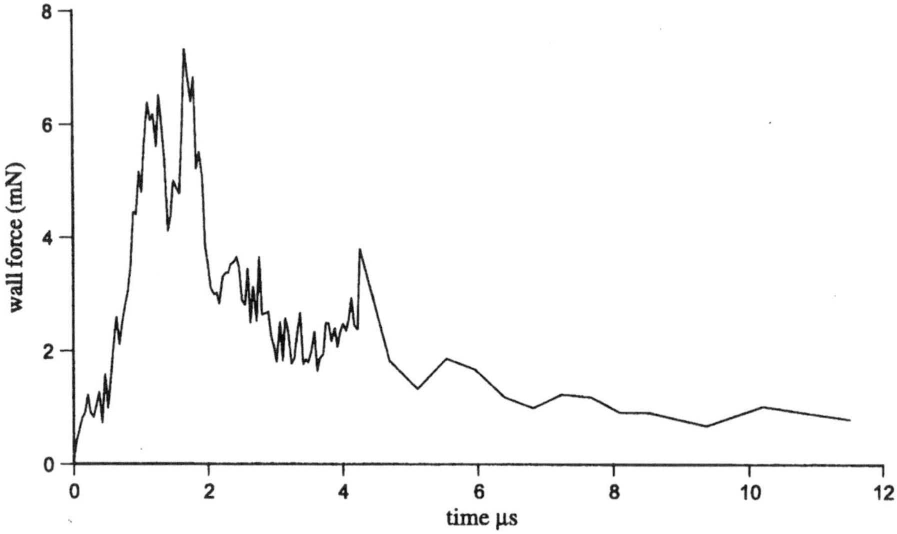

Fig. 6.27 Evolution of the total normal wall force

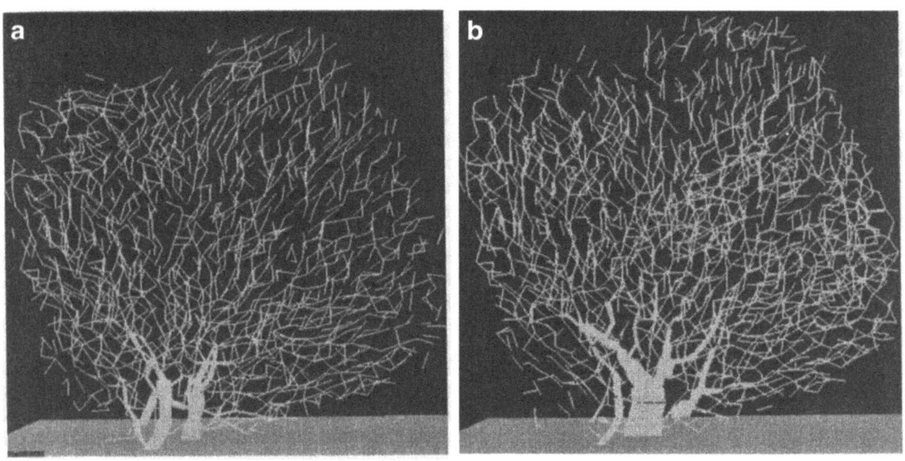

Fig. 6.28 Force transmission (**a**) after 1.125 μs (**b**) after 1.892 μs

maximum value of 7.3 mN after 1.64 μs. There is then a sudden drop in the force to about 3 mN after which the force reduces further with significant fluctuations until, after 65 μs, the force is approximately 1 μN corresponding to the self-weight of the agglomerate.

Figure 6.28 illustrates the force transmission through the agglomerate. The lines show the location and orientation of the (resultant) contact forces. The thickness of

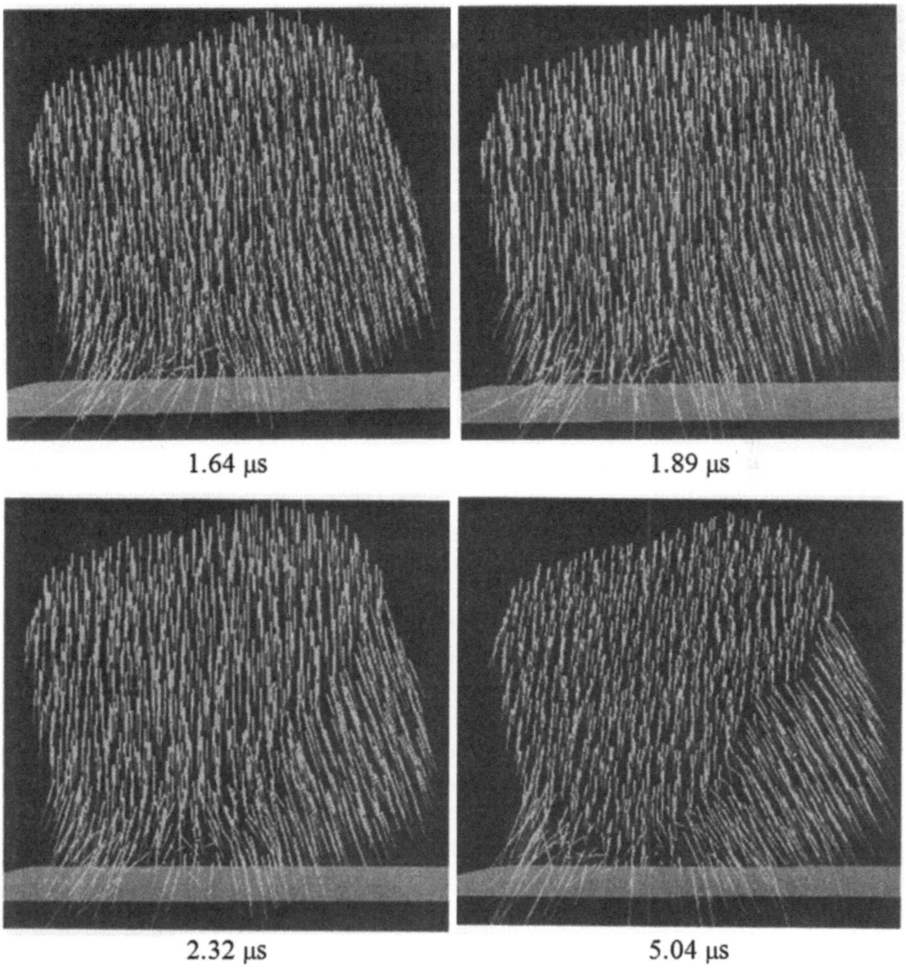

Fig. 6.29 Particle velocity fields at various elapsed times

the lines indicates the magnitude of the force scaled to the current maximum. For clarity, only the thin central section seen in Fig. 6.26b is shown. Figure 6.28a shows that, when the wall force is 6.5 mN, the large forces generated at the contacts with the wall propagate vertically upwards. Figure 6.28b shows that, just after the wall force has reached its maximum value, there is a significant but smaller wall force propagating towards the lower right-hand side of the agglomerate.

Figure 6.29 shows the particle velocity field as it evolves during the impact. It can be seen that, as a consequence of the large forces transmitted vertically upwards into the agglomerate, the primary particles in the region into which these forces propagate are decelerated but continue to move downwards in a vertical direction. Particles in the lower right-hand side of the agglomerate do not experience such a rapid deceleration and, consequently, a heterogeneous velocity field is created, which results in an inclined velocity discontinuity between the loaded and unloaded regions. The relative shear motion along the velocity discontinuity causes some breakage of contacts and thereby a weakened plane is created. As a result of the secondary, inclined contact force transmission seen in Fig. 6.28b, there is a rotation of the velocity field in the lower right-hand region of the agglomerate that increases the shear weakening along the velocity discontinuity and results in the fracture plane shown in Fig. 6.26b. Further details are provided by Thornton and Liu (2004).

6.2 Agglomerate-Agglomerate Collisions

One would expect that, for comparable impact speeds, agglomerate-agglomerate collisions would be less destructive than agglomerate-wall collisions. In this section we briefly consider agglomerate-agglomerate collisions.

Figure 6.30 illustrates the breakage resulting from oblique collisions between two identical spherical agglomerates when the relative impact speed is 1 m/s. Each agglomerate consists of a polydisperse system of 2000 primary particles which were modelled as autoadhesive, elast-plastic spheres, see Sect. 3.4 for details. The average diameter of the primary particles was 20 μm and the nominal size of the agglomerates was 0.176 mm. The interface energy $\Gamma = 2$ J/m^2. A timestep of 9 ns was used and the collision duration varied from 3 to 9 ms.

The two agglomerates are identical but the second agglomerate is simply a copy of the first agglomerate that was translated to provide a small initial gap between the two agglomerates. Hence, the microstructures adjacent to the point of impact were different and, therefore, the force transmission through the two agglomerates was different. Consequently, as can be seen in Fig. 6.30, the two agglomerates experienced different breakage patterns.

The figure illustrates the effect of impact angle on the amount of breakage that occurred when the impact angle was varied between 0 and 45°. For impacts at angles greater than 45° the breakage was limited to the production of singlets that were abraded from the agglomerates with the number of singlets reducing with increase in the impact angle. To quantify this, Fig. 6.31 shows how the damage ratio varies with impact angle for a strong ($\Gamma = 2.0$ J/m^2) and a weak ($\Gamma = 0.2$ J/m^2) agglomerate.

In Fig. 6.32a, the damage ratio, D, is replotted against a Weber number, W, which is defined as

$$W = \frac{\rho \ d_p V_n^2}{\Gamma} \tag{6.7}$$

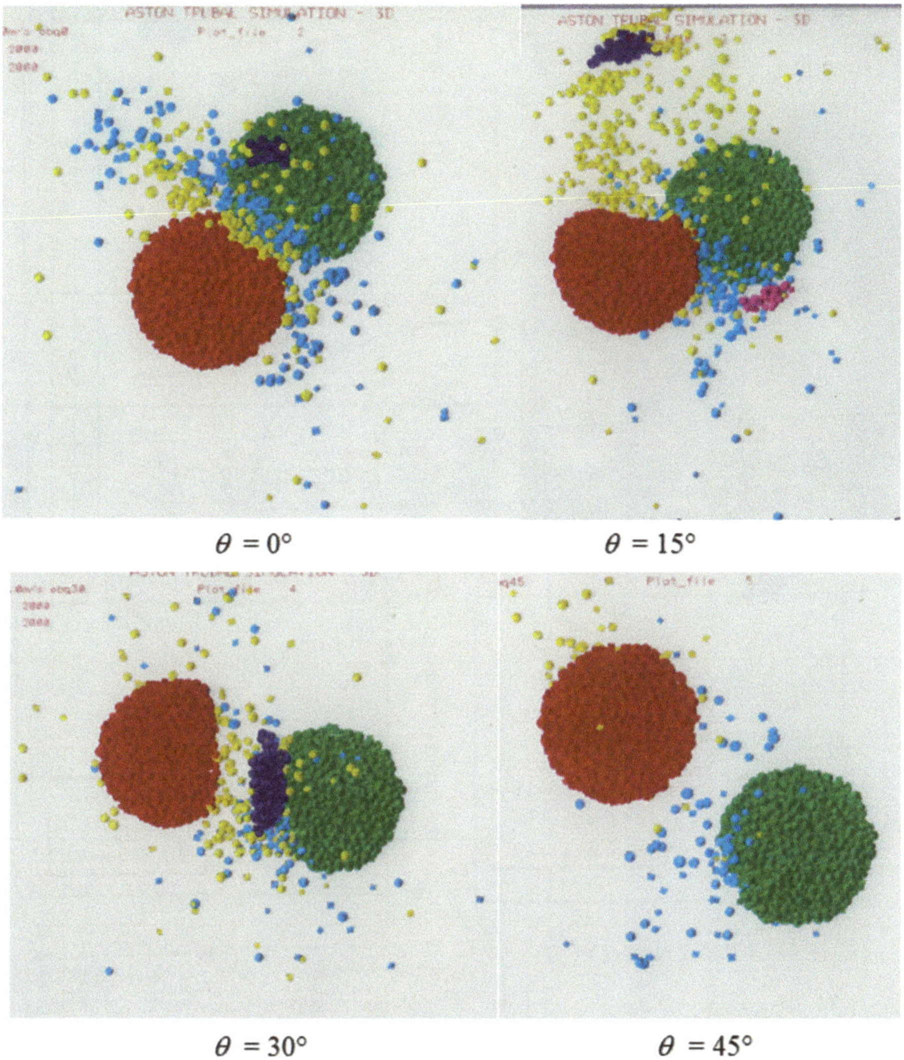

Fig. 6.30 Oblique agglomerate-agglomerate collisions (relative impact speed 1.0 m/s)

where ρ and d_p are the solid density and mean primary particle size and V_n is the normal component of the relative impact speed. From the figure, the power law scaling is $D = 0.86\,W^{0.78}$. Defining the debris as being composed of fragments consisting of less than 20 primary particles, it is found that the mass fraction of debris produced is linearly proportional to the Weber number, as shown in Fig. 6.32b.

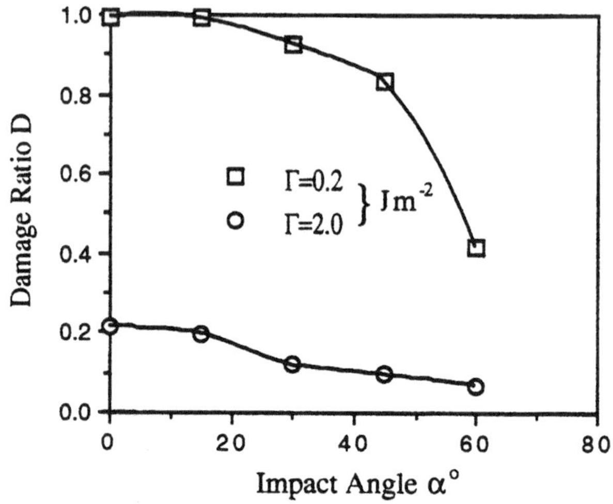

Fig. 6.31 Variation of damage ratio with impact angle (relative impact speed = 1 m/s)

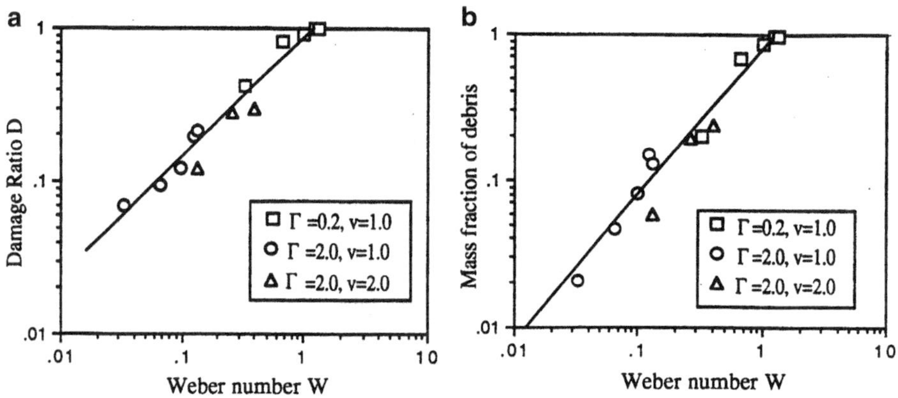

Fig. 6.32 Dependency on the Weber number of (**a**) damage ratio (**b**) mass fraction of debris

References

Arbiter, N., Harris, C.C., Stamboltzis, G.A.: Single fracture of brittle spheres. Trans. Soc. Min. Eng. AIME **244**, 118–133 (1969)

Kafui, K.D., Thornton, C.: Numerical simulations of impact breakage of a spherical crystalline agglomerate. Powder Technol. **109**, 113–132 (2000)

Liu, L., Kafui, K.D., Thornton, C.: Impact breakage of spherical, cuboidal and cylindrical agglomerates. Powder Technol. **199**, 189–196 (2010)

Mishra, B.K., Thornton, C.: Impact breakage of particle agglomerates. Int. J. Miner. Process. **61**, 225–239 (2001)

Moreno, R., Ghadiri, M., Antony, S.J.: Effect of impact angle on the breakage of agglomerates: a numerical study using DEM. Powder Technol. **130**, 132–137 (2003)

Takagi, Y., Mizutani, H., Kawakami, S.: Impact fragmentation experiment of basalt and pyrophyllites. Icarus **59**, 462–477 (1984)
Thornton, C., Liu, L.: How do agglomerates break? Powder Technol. **143–144**, 110–116 (2004)
Thornton, C., Yin, K.K., Adams, M.J.: Numerical simulation of the impact fracture and fragmentation of agglomerates. J. Phys. D. Appl. Phys. **29**, 424–435 (1996)
Thornton, C., Ciomocos, M.T., Adams, M.J.: Numerical simulations of agglomerate impact breakage. Powder Technol. **105**, 74–82 (1999)
Yin, K.K.: Numerical modelling of agglomerate degradation. PhD thesis, Aston University (1992)

Chapter 7
Fluidised Beds

Abstract Gas-fluidised beds have been extensively studied in academia and widely used in industry. When examined at the macro-scale, fluidised beds may appear to behave like a solid, a liquid or a gas, depending on the magnitude of the applied superficial gas velocity. One of the attractions of DEM is that it can model all three different phases. To account for the interstitial gas, a combined Lagrangian-Eulerian approach is used by combining the use of DEM for the particle phase with CFD modelling of the gas phase. In this chapter, following an initial outline of the theoretical background to the methodology, results of 2D and 3D simulations of fluidized beds are presented. Using 2D simulations, we examine the different types of behaviour as the gas velocity is increased to cover the complete range from fixed bed to homogeneous expansion, bubbling, turbulent and fast fluidisation. we also examine how the transition from fixed bed to bubbling bed, i.e. the so-called homogeneous expansion regime, is affected when surface energy is attributed to the particles. Then, using 3D simulations, we examine and provide visualisations of bubble formation, bubble rise and bubble splitting.

In this chapter the results of simulations of fluidised beds are presented. This is restricted to the author's own work that was motivated by the need to re-examine historical issues that had not been completely resolved. For more general applications of fluidised bed simulations the reader is referred to papers published by Prof. Tsuji's group in Osaka, Prof. Yu's group at Monash (previously UNSW) and the group of Prof. Kuipers at Eindhoven (previously Twente).

The increasing power of computer hardware has made a Lagrangian-Eulerian modelling of gas-solid fluidisation feasible following the original work of Tsuji et al. (1993) who combined the discrete element method (DEM) modelling of the particle phase with computational fluid dynamics (CFD) modelling of the fluid phase to simulate fluidised beds in 2D. This combined DEM-CFD approach is fully described by Kafui et al. (2002).

In this chapter the results of simulations of fluidised beds are presented. This is restricted to the author's own work that was motivated by the need to re-examine historical issues that had not been completely resolved. For more general applications of fluidised bed simulations the reader is referred to papers published by Prof. Tsuji's group in Osaka, Prof. Yu's group at Monash (previously UNSW) and the group of Prof. Kuipers at Eindhoven (previously Twente).

C. Thornton, *Granular Dynamics, Contact Mechanics and Particle System Simulations*, Particle Technology Series 24, DOI 10.1007/978-3-319-18711-2_7

7.1 Theoretical Considerations

The total force F_i acting on particle i in a fluidised bed has a number of components: a gravitational force $m_i g$, a fluid-particle interaction force F_{fpi} and a solid particle-particle contact force F_{ci}. The translational and rotational motions of each particle are governed by the equations

$$m_i \frac{d\ v_i}{d\ t} = F_i = F_{ci} + F_{f\,pi} + m_i g \tag{7.1}$$

$$I_i \frac{d\ \omega_i}{d\ t} = T_i \tag{7.2}$$

in which T_i is the torque arising from the tangential components of the contact force and I_i, v_i and ω_i are the moment of inertia, linear velocity and angular velocity of the particle respectively. Equations (7.1) and (7.2) correspond to Eqs. (2.1) and (2.2) but with the fluid-particle interaction force F_{fpi} added in Eq. (7.1).

By numerically integrating Eqs. (7.1) and (7.2) twice, using Eqs. (2.3) and (2.4), new velocities and positions of the particles are obtained and from the new positions and velocities of the particles new fluid-particle interaction forces and solid-solid contact forces can be calculated.

7.1.1 Fluid-Particle Interaction Force

Following Anderson and Jackson (1967), the force exerted by the fluid on each particle, F_{fpi}, can be written as the sum of a component due to ‘macroscopic’ variations in the fluid stress tensor σ_f and a component $F^*_{f\,pi}$ due to detailed variations of the point stress tensor in the fluid flow field around a particle.

$$F_{f\,pi} = V_{pi} \nabla \cdot \sigma_f + F^*_{f\,pi} \tag{7.3}$$

where V_{pi} is the volume of the particle. The local average stress tensor in the fluid may be written as

$$\sigma_f = -p\delta + \tau_f \tag{7.4}$$

where p is the fluid pressure, δ is the identity tensor and τ_f is the deviatoric stress tensor. Assuming a Newtonian fluid with a viscous stress tensor dependent only on the fluid motion, following Bird et al. (1960), we may write

$$\tau_f = \left[\left(\mu_b - \frac{2}{3}\mu_s\right) \nabla \cdot u\right] \delta + \mu_s \left[(\nabla u) + (\nabla u)^{-1}\right] \tag{7.5}$$

where u, μ_b and μ_s are the velocity, bulk viscosity and shear viscosity of the fluid respectively.

The second term on the right of Eq. (7.3) includes skin friction and drag contributions and is made up of (i) an effective drag force in the direction of the relative velocity between the fluid and the particle and (ii) a virtual or added mass force accounting for the resistance of the fluid mass that is moving at the same acceleration as the particle. In gas-solid systems the virtual mass term is negligible and F^{*}_{fpi} reduces to the effective drag force which, according to Anderson and Jackson (1967), is the drag force F_{di} obtained from the experimentally based correlations multiplied by the local void fraction ε. Consequently, substituting Eq. (7.4) into Eq. (7.3) the fluid-particle interaction force is given by

$$F_{fpi} = -V_{pi}\nabla \cdot p + V_{pi}\nabla \cdot \tau_f + \varepsilon F_{di} \tag{7.6}$$

The drag force F_{di} is calculated using the following empirical correlation of Di Felice (1994) which provides a continuous variation of drag force over the full practical range of flow regimes and voidages.

$$F_{di} = \frac{1}{2} C_{Di}\rho_f \frac{\pi d_{pi}^2}{4} \varepsilon_j^2 |u_j - v_i| (u_j - v_i) \varepsilon_j^{-(\chi+1)} \tag{7.7}$$

where ρ_f is the fluid density, d_{pi} is the particle diameter and the subscript j for the fluid velocity u and the voidage ε denotes the computational fluid cell in which particle i resides. The fluid drag coefficient for a single, unhindered particle C_{Di} is calculated from

$$C_{Di} = \left[0.63 + \frac{4.8}{\sqrt{Re_{pi}}}\right]^2 \tag{7.8}$$

and the particle Reynolds number Re_{pi} is based on the superficial slip velocity between particle and fluid

$$Re_{pi} = \frac{\rho_f d_{pi} \varepsilon_j |u - v|}{\mu_s} \tag{7.9}$$

The term $\varepsilon_j^{-(\chi+1)}$ in Eq. (7.7) is a correction for the presence of other particles with

$$\chi = 3.7 - 0.65 \ \exp\left[-\frac{(1.5 - \log_{10} Re_{pi})^2}{2}\right] \tag{7.10}$$

7.1.2 Particle-Fluid Interaction Force

For the fluid hydrodynamics, the continuity and momentum equations used in the PGF model of Kafui et al. (2002) are

$$\frac{\partial(\varepsilon\rho_f)}{\partial t} + \nabla\cdot(\varepsilon\rho_f u) = 0 \tag{7.11}$$

$$\frac{\partial(\varepsilon\rho_f u)}{\partial t} + \nabla\cdot(\varepsilon\rho_f uu) = -\nabla p + \nabla\cdot\tau_f - F_{pf} + \varepsilon\rho_f g \tag{7.12}$$

in which the particle-fluid interaction force F_{pf} is obtained by summing up the fluid-particle interaction forces F_{fpi} acting on all the particles in a fluid cell n_c and dividing by the volume of the fluid cell ΔV_c.

$$F_{pf} = \frac{\sum_{i=1}^{n_c} F_{fpi}}{\Delta V_c} \tag{7.13}$$

The ideal gas law is used to calculate the fluid density ρ_f.

$$\rho_f \frac{M_f}{RT} p \tag{7.14}$$

where the average molecular weight of air $M_f = 0.0288$ kg/mol at a temperature $T = 293$ K.

7.2 2D Simulations

Gas-fluidised beds have been extensively studied in academia and widely used in industry. When examined at the macro-scale, fluidised beds may appear to behave like a solid, a liquid or a gas, depending on the magnitude of the applied superficial gas velocity. These "phase transitions" are important since the fundamental rate parameters for reactor design and operation change in accordance with the flow regimes of fluidisation. One of the attractions of DEM-CFD modelling is that it can model all three different phases.

A series of 2D DEM-CFD simulations of fluidised bed behaviour was performed, Thornton et al. (2015), to examine the transitions from fixed bed to homogeneous expansion, bubbling, turbulent and fast fluidisation. Figure 7.1 illustrates how the bed behaviour changes as the superficial gas velocity is increased. The bed width corresponds to 40 times the mean particle diameter and, as the gas velocity was increased, the height of the container was adjusted to ensure that no particles reached the top of the container. The whole of the container was divided into small square computational fluid cells of dimension five times the mean particle diameter. A polydisperse system of 5000 elastic spheres was used. All the particles were initially randomly generated as a granular gas (no contacts) inside the container with all the particle centres located in the same plane and subsequent out-of-plane motion was suppressed. A vertical gravity field was then introduced in order to create a pluvially deposited bed of particles. The mean particle diameter

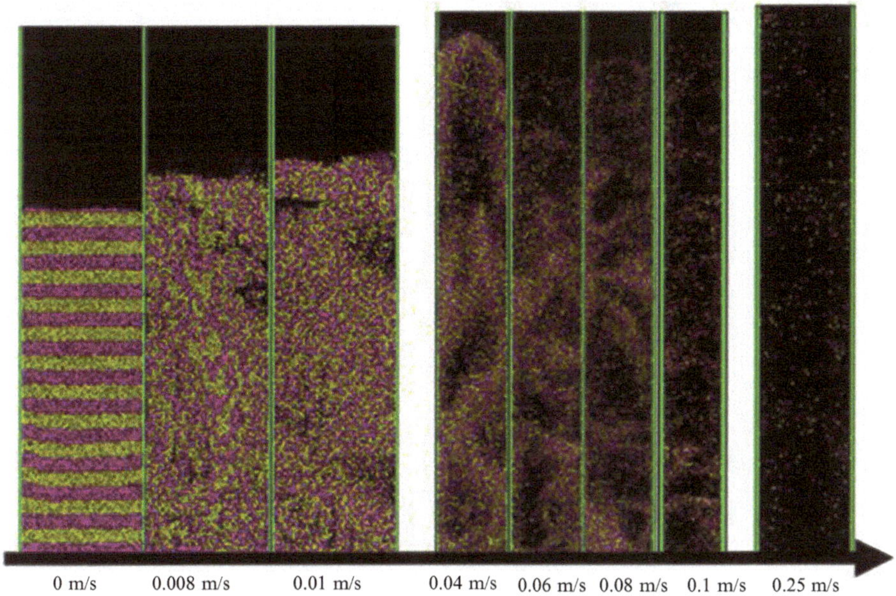

Fig. 7.1 Typical particle configurations for different superficial gas velocities

was 50 μm, the initial bed height was 6.54 mm with an initial voidage of 0.459 and 8,465 interparticle contacts. Further simulation details are provided by Thornton et al. (2015).

An initial uniform gas velocity $U = 0.0003$ m/s was introduced into the bed from the bottom row of computational fluid cells. The pressure drop across the bed was obtained as the time-averaged difference between the average pressure in the bottom and top rows of fluid cells. This was repeated for a range of gas velocities incremented in relatively small steps up to $U = 1.2$ m/s. With increasing gas velocity, as seen in Fig. 7.1, bed expansion increases and the particles are transported higher. Figure 7.2 shows how the average void fraction of the bed changes with increasing superficial gas velocity. There are clearly three regimes. At low gas velocities the void fraction does not change. This corresponds to the fixed bed regime that exhibits solid-like behaviour. At high gas velocities, as the void fraction $\varepsilon \to 1$, the behaviour is gas-like corresponding to fast fluidisation, as in the riser of a circulating fluidised bed. Between these two regimes the bed behaves like a liquid but there is no indication in the figure to distinguish between different sub-regimes. In Fig. 7.1, however, it can be seen that there are three sub-regimes corresponding to homogeneous expansion, bubbling fluidisation and turbulent fluidisation. The transitions between these sub-regimes are considered below.

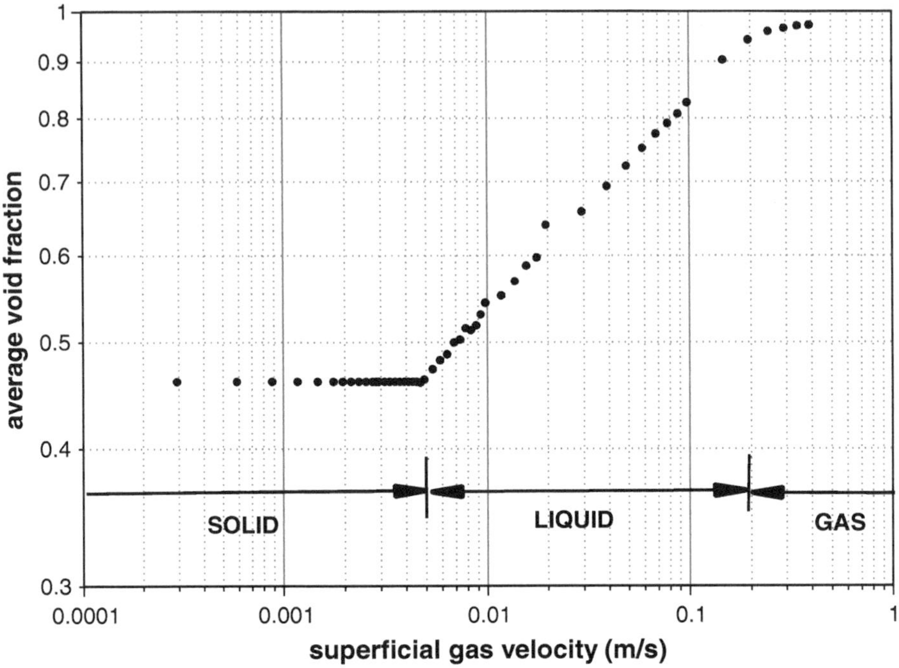

Fig. 7.2 Variation of the average bed void fraction with increasing superficial gas velocity

7.2.1 The Transition from Fixed to Bubbling Bed

Conventionally, the point when the average pressure drop first becomes equal to the bed weight divided by the cross-sectional area of the bed is defined as 'minimum fluidisation' and the gas velocity at which this occurs is denoted as U_{mf}. Above U_{mf} the pressure drop remains constant and bed expansion occurs. In Fig. 7.3, the average pressure drop is normalised by dividing by the bed weight per unit area. From the figure, $U_{mf} = 0.0048$ m/s, which is in reasonable agreement with the value of 0.0041 m/s predicted using the Ergun (1952) correlation.

Superimposed in Fig. 7.3 is the number of interparticle contacts normalised by the initial number of contacts when the bed had been deposited. It can be seen that some contacts were broken, without any significant change in voidage/bed height, prior to minimum fluidisation and that, above U_{mf}, the average number of contacts decreased at a decreasing rate until an asymptotic value of about 5 % of the initial number of contacts was reached when $U = 0.01$ m/s. Also superimposed in the figure is the mechanical coordination number Z_m defined by Eq. (2.22). From the figure it can be seen that, when $U = U_{mf}$, $Z_m = 3$ which, in 2D, corresponds to an isostatic state, see Sect. 2.2.1.

Figure 7.4 shows the expansion of the bed as the gas velocity is increased to 0.01 m/s. From examination of video sequences of the simulations it was observed

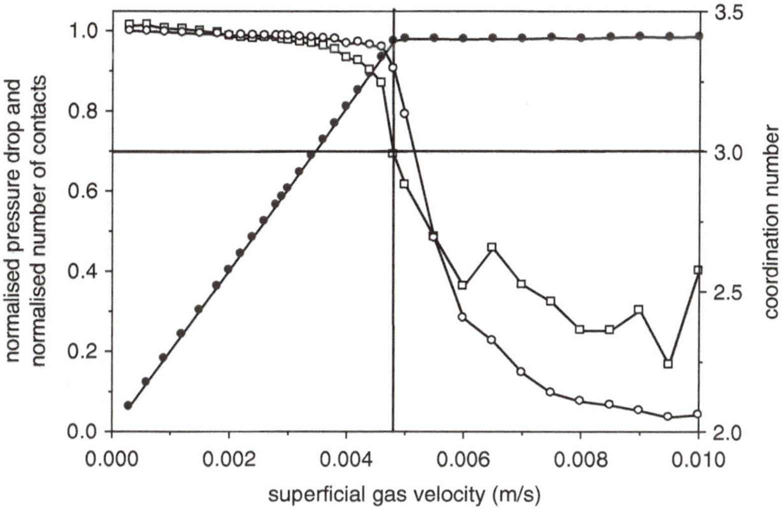

Fig. 7.3 Normalised pressure drop (*solid circles*), normalised number of contacts (*open circles*) and mechanical coordination number (*open squares*)

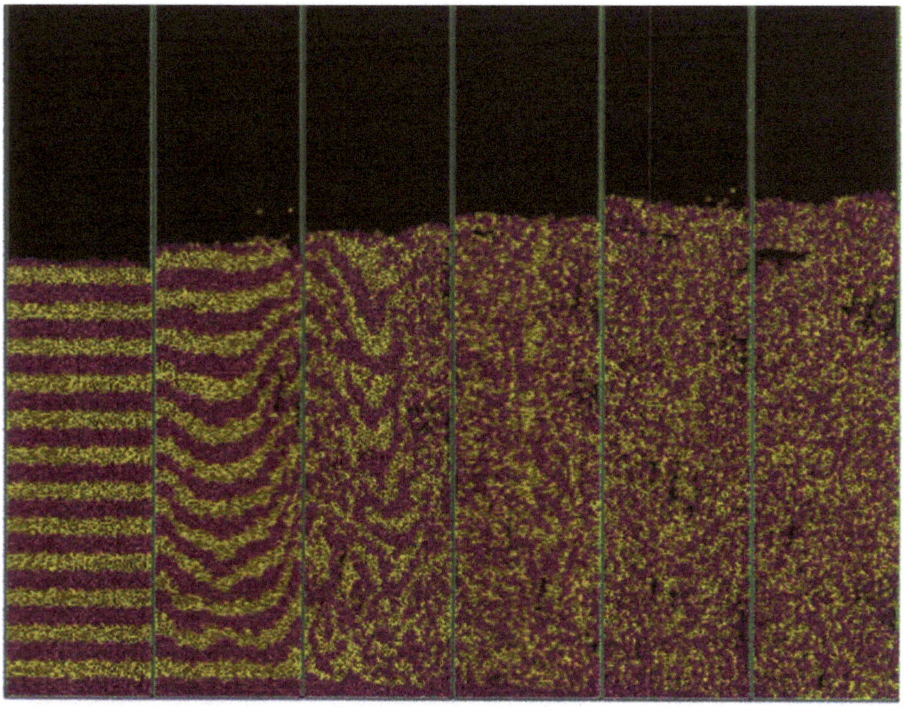

u = 0.0046[m/s] u = 0.0060[m/s] u = 0.0075[m/s] u = 0.0085[m/s] u = 0.0095[m/s] u = 0.01[m/s]

Fig. 7.4 Bed expansion with increasing gas velocity

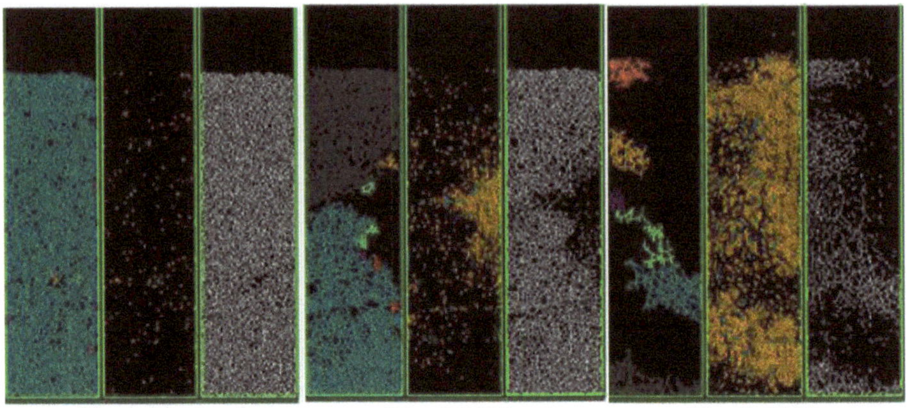

Fig. 7.5 Cluster visualisation at the start of the 'homogeneous expansion' regime

that the first bubble eruption at the bed surface occurred when $U = 0.01$ m/s, which therefore corresponds to what is termed the minimum bubbling velocity, U_{mb}.

The gas velocity range $U_{mf} < U < U_{mb}$ is conventionally known as the homogeneous expansion regime. Figure 7.5 shows snapshots to illustrate the evolution of the structure of the bed at the start of the bed expansion. In each snapshot, the three snapshots show (i) the six largest cluster sizes in the system (left column), (ii) singlets, doublets and triplets (centre column) and (iii) the spatial distribution of interparticle contacts (right column). Note that clusters of intermediate size are not shown. It can be seen that the number of contacts decreases sharply for 0.0048 m/s $< U <$0.006 m/s with a corresponding sharp increase in the number of fines. The figure clearly shows the degradation of large clusters, the increase in fines production and the corresponding loss of contacts as the gas velocity increases.

From Fig. 7.5 it is also clear that, at least at the start of the "homogeneous expansion" regime the bed is not in fact homogeneous. The results of the simulations suggest that the so-called homogeneous expansion regime is actually a transition regime. At U_{mf} the bed is at an isostatic state that is the start of a transition from solid-like to fluid-like behaviour and that only when the contact number reaches a small asymptotic value is the bed 'fully fluidised' and bubbling can begin.

7.2.2 The Transition from Bubbling Bed to Turbulent Bed

Above U_{mb} bubbling occurs, with the size of the bubbles increasing with increase in gas velocity. As a consequence of bubble eruption at the bed surface the amplitude of the pressure drop fluctuations also increases with increase in gas velocity. In the bubbling regime both bubble splitting and bubble coalescence occur. When bubble

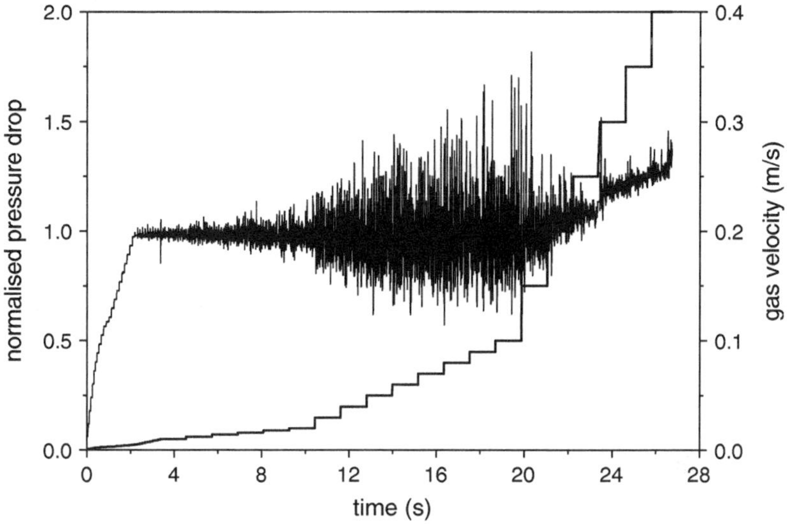

Fig. 7.6 Non- averaged pressure drop fluctuations

splitting exceeds bubble coalescence the mean bubble size decreases leading to a decrease in the amplitude of the pressure drop fluctuations, as shown in Fig. 7.6.

Figure 7.7 shows how the standard deviation of the normalised pressure drop increases to a maximum value and then reduces at a decreasing rate towards an asymptotic value. Yerushalmi et al. (1978) and Yerushalmi and Cankurt (1979) suggested that the gas velocity U_c at which the standard deviation of the pressure drop reaches a maximum value indicates the beginning of a transition to turbulent fluidisation. They also suggested that the gas velocity U_k at which the standard deviation of the pressure drop levels off at some low value indicates the end of the transition. However, subsequent researchers have adopted U_c as the start of the turbulent regime and U_k as the end of the turbulent regime and the transition to fast fluidisation (Bi et al. (2000). From Fig. 7.7, $U_c = 0.085$ m/s but it is clear from the figure that the above definition of U_k is ambiguous. Figure 7.7 also shows that the simulation results indicate that in the turbulent regime the average pressure drop increases with increasing gas velocity.

7.2.3 The Transition from Turbulent to Fast Fluidisation

The voidage data shown in Fig. 7.2 for the 'liquid' regimes indicates power law behaviour. This is confirmed in Fig. 7.8 in which the void fraction and the gas velocity have been normalised by the corresponding values at minimum fluidisation, i.e. $\varepsilon_{mf} = 0.459$ and $U_{mf} = 0.0048$ m/s respectively. From the best fit line

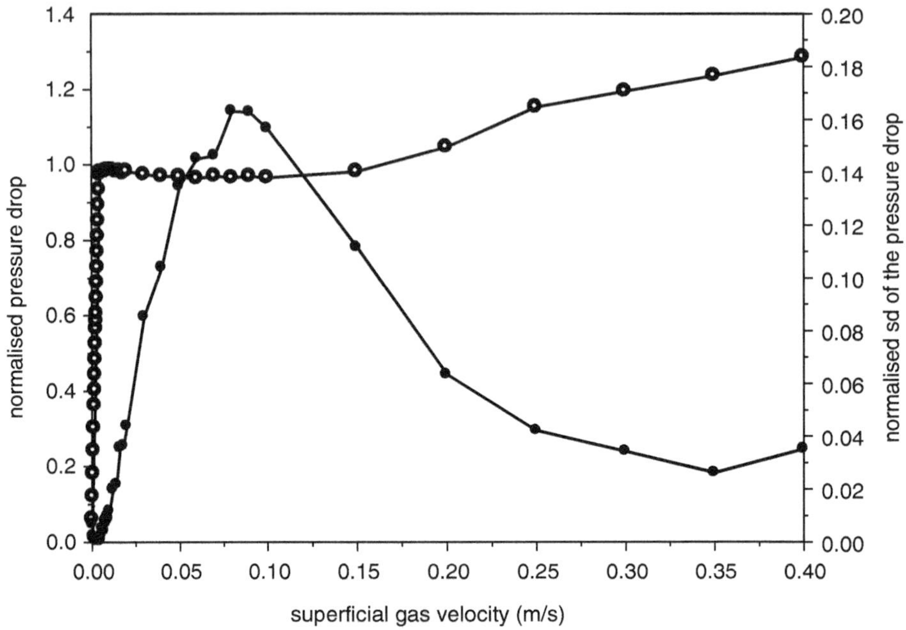

Fig. 7.7 Variation of the standard deviation of the normalised pressure drop with superficial gas velocity

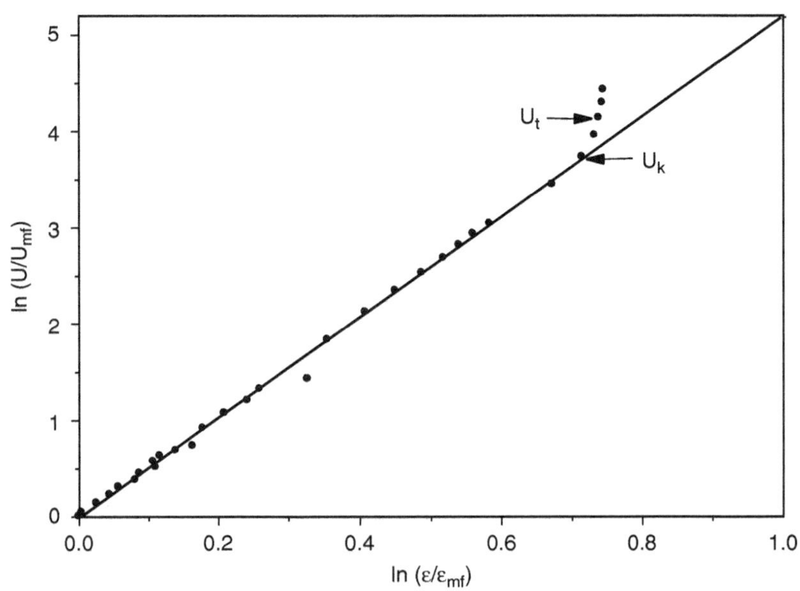

Fig. 7.8 Power law fit to the data in the bubbling and turbulent regimes

$$\frac{U}{U_{mf}} = \left(\frac{\varepsilon}{\varepsilon_{mf}}\right)^{5.2} \tag{7.15}$$

and therefore

$$U = 0.275\varepsilon^{5.2} \tag{7.16}$$

Equation (7.16) implies that when $\varepsilon = 1$ the gas velocity of 0.275 m/s corresponds to the free-fall terminal velocity U_t of an isolated sphere in an infinite fluid. However, using a single average-sized sphere $d_p = 50$ μm and atmospheric gas with the initial gas velocity set to zero, the gravity driven free-fall of an isolated particle was simulated and the terminal velocity was found to be 0.3 m/s. The data point corresponding to U_t is indicated on Fig. 7.8.

Godard and Richardson (1969) proposed an alternative expression for the exponent n

$$n = \frac{log(U_{mf}/U_t)}{log\varepsilon_{mf}} \tag{7.17}$$

from which, using $U_t = 0.3$ m/s, gives $n = 5.266$ which is very close to the value of 5.2 in Eq. (7.16).

It can be seen from Fig. 7.8 that the data points deviate from the power law relationship given by Eq. (7.15) at very high gas velocities in the fast fluidisation regime. This suggests that a rational definition of the transition from turbulent fluidisation to fast fluidisation is provided by the gas velocity at which the data first deviates from Eq. (7.15) and on this basis $U_k = 0.2$ m/s, as indicated on Fig. 7.8.

7.2.4 *Effect of Surface Energy*

In the above simulations the average particle size was 50 μm and at this size the particles would be expected to be adhesive due to van der Waals forces. Consequently a series of simulations was performed on the same particle system to examine the effect of surface energy on the fluidised bed behaviour (Yang et al. 2013) using the adhesive-elastic contact force model described in Sect. 3.3. In the JKR model of adhesive-elastic particle interactions, the maximum tensile force required to break a contact is given by Eq. (3.71) from which, for a polydisperse system, there will be a range of pull-off forces F_{nc} depending on the radii of the two particles in contact. Consequently, values of interface energy Γ are selected to make the average bond strength a multiple of the average particle weight.

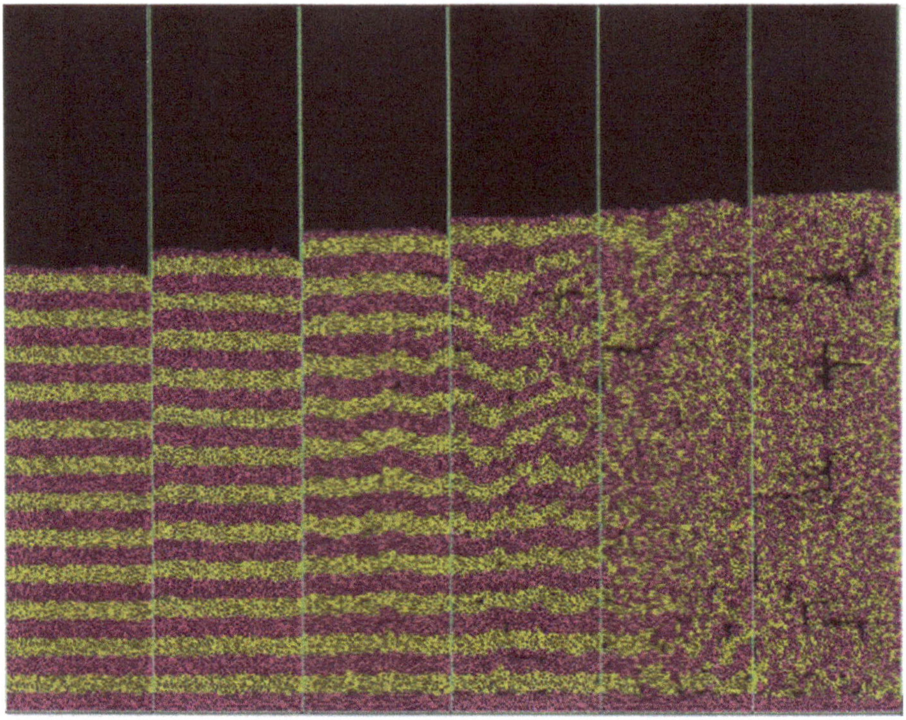

Fig. 7.9 Bed expansion with increasing gas velocity ($K = 1$)

$$\langle F_{nc} \rangle = K \langle mg \rangle \tag{7.18}$$

For example, for $K = 1$ and interface energy $\Gamma = 27.25$ μJ/m^2 the pull-off forces F_{nc} are in the range 1.605 nN ±10 %.

This series of simulations focussed on the transition from fixed bed to bubbling bed, i.e. the so-called 'homogeneous expansion' regime. Figure 7.9 shows the bed expansion for $K = 1$, which can be compared with the $K = 0$ case illustrated in Fig. 7.4.

Figure 7.10 provides comparisons between typical snapshots of the particle configurations for different surface energies when the gas velocity was 0.01 m/s and 0.02 m/s. When $K = 1$ and $K = 2$ the figure shows that, with a gas velocity $U = 0.01$ m/s, the bed has expanded without any obvious bubble observed. The figure also shows that, for $K = 5$, channelling occurs when $U = 0.01$ m/s. However, when the gas velocity is increased to $U = 0.02$ m/s, fully developed bubbles are observed in all cases. This suggests that for $0 \leq K \leq 5$ the minimum bubbling velocity lies in the range 0.01 m/s $\leq U_{mb} \leq$ 0.02 m/s and that with increasing surface energy a higher gas velocity is required to reach the bubbling regime, if that is possible.

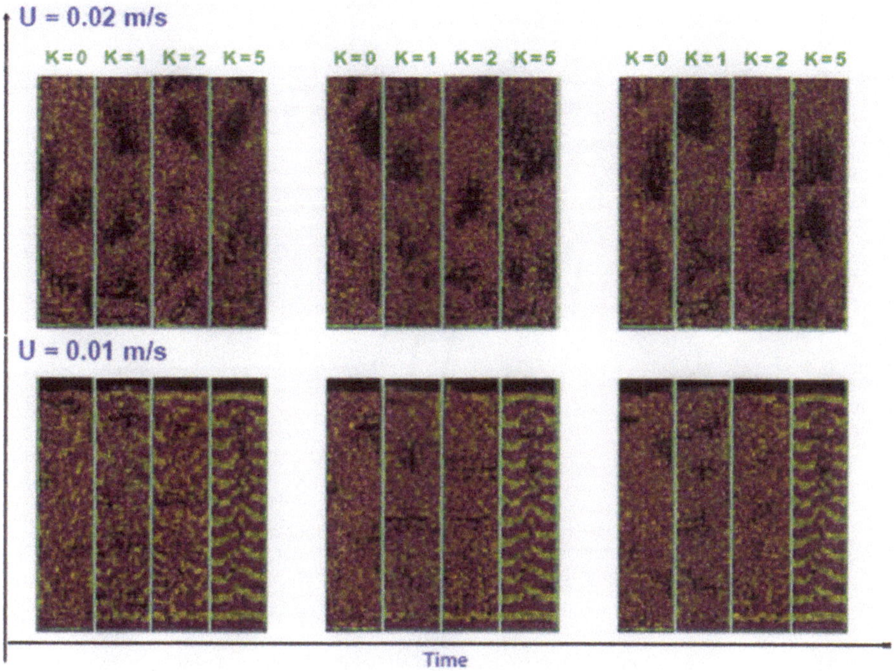

Fig. 7.10 Typical snapshots of the bed for gas velocities of 0.01 m/s and 0.02 m/s and $K = 0$, 1, 2 and 5

In Fig. 7.11, the normalised pressure drop and the normalised number of contacts are plotted against the superficial gas velocity. It can be seen that in the fixed bed regime the pressure drop curve is independent of surface energy. However, in the simulations, the surface energy was introduced after pluvial deposition of the particle bed. It would have been more realistic to introduce surface energy prior to pluvial deposition. This in itself would lead to a higher bed voidage and consequently U_{mf} would increase with increase in surface energy. Nevertheless, the results shown in Fig. 7.11 demonstrate that any increase in U_{mf} due to surface energy is solely due to the higher voidage and not to any increased bed resistance resulting from stronger interparticle bonds.

Figure 7.11 also shows that a pressure drop overshoot occurs for systems with surface energy. For $K = 1$ and $K = 2$ the overshoot is slight but, for $K = 5$, the average pressure drop increases to a maximum value that is 10 % greater than that necessary to balance the self-weight of the bed. To understand the overshoot phenomenon, a bed with vertical periodic boundaries was simulated. The results, shown later in Fig. 7.15, indicate that the overshoot is solely due to the wall effect and, for all cases, minimum fluidisation occurs when the normalised pressure drop first equals unity, which occurs before any overshoot appears.

The most notable aspect of Fig. 7.11 is that, for $K > 1$, bond breakage occurs in two stages: (i) approximately 40 % of the bonds are broken, creating a 'partially

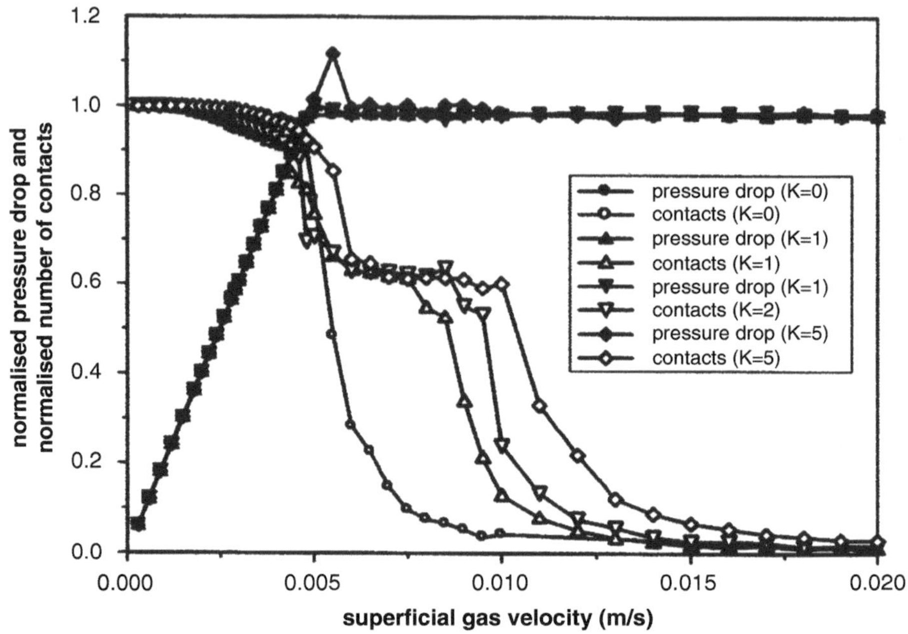

Fig. 7.11 Effect of surface energy on the evolution of pressure drop and number of contacts

fluidised' bed, and (ii) for a higher bond strength a higher gas velocity is required to break the remaining bonds in order to 'fully fluidise' the bed. When the gas velocity is increased above U_{mf} a sufficient number of contacts is broken to form a few relatively large agglomerates that consist of most of the particles in the bed. The strength of the agglomerates so formed depends on the strength of the interparticle bonds. As shown in Fig. 7.11, in order to fracture the initially formed agglomerates a higher gas velocity is required for higher values of surface energy. Once this has been achieved, further disintegration of the agglomerates continues progressively until the fines predominate and occupy the whole bed. At this point the bed can be considered to be 'fully fluidised' and bubbling can occur. The process is illustrated in Fig. 7.12 for the case of $K = 1$.

For each gas velocity, the three columns show (i) the six largest agglomerates (left), (ii) singlets, doublets and triplets (centre) and (iii) the spatial distribution of interparticle contacts (right); agglomerates of intermediate sizes are not shown. Initially, the structure of the bed breaks into two once U_{mf} is reached. The amount of fines remains low until $U = 0.0075$ m/s. At this point, the second large reduction in the number of contacts begins, see Fig. 7.11, when the gas velocity is sufficiently high to break up the large agglomerates, leading to a corresponding rapid increase in the number of fines particles. The connectivity network completely disintegrates until finally the bed consists entirely of singlets, doublets and triplets and is fully fluidised and bubbling begins when $U = 0.014$ m/s. Typical snapshots of the beds at minimum bubbling velocity are shown in Fig. 7.13 for $K = 0$, 1, 2 and 5.

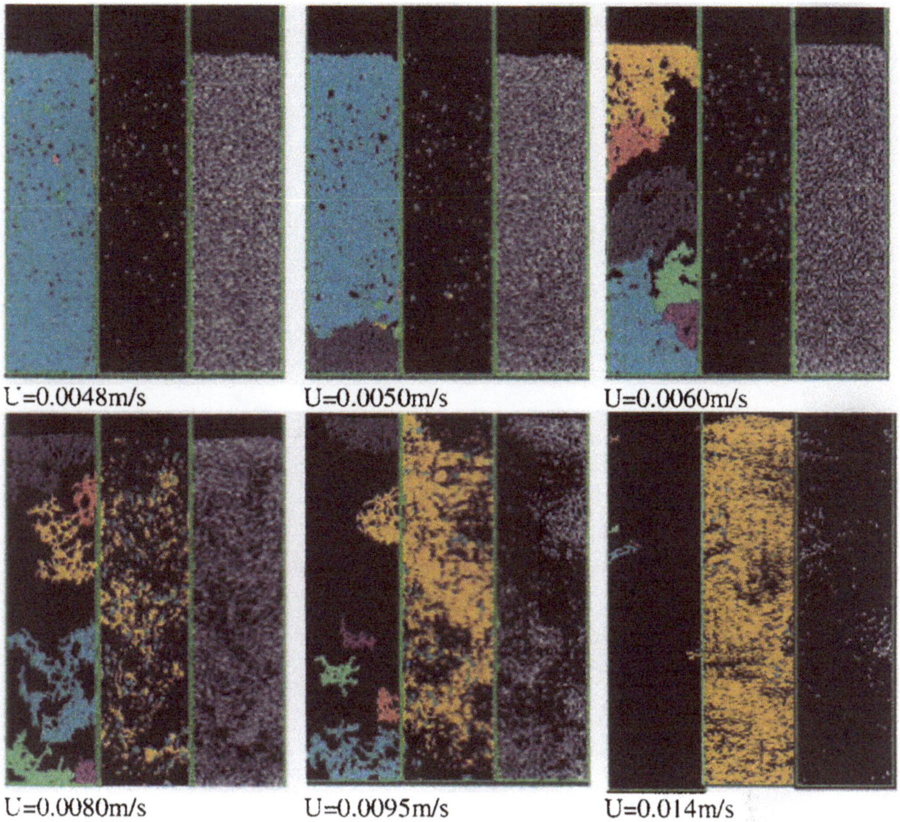

Fig. 7.12 Agglomerate degradation during 'homogeneous expansion'

The average pressure drop and average mechanical coordination number are plotted against superficial gas velocity, for different values of K, in Fig. 7.14. It can be seen that the average mechanical coordination number, defined by Eq. (2.22), is approximately 3 when U reaches U_{mf}, except for the case of $K = 5$. For the case of $K = 5$, the data is replotted in Fig. 7.15. The figure shows results obtained for both a wall bounded system and a system with vertical periodic boundaries. It can be seen that, when vertical periodic boundaries are used, there is no overshoot in the pressure drop. This indicates that the pressure drop overshoot is an artefact resulting from the extra kinematic constraint provided by the wall boundaries.

It can be seen that, for the case of vertical periodic boundaries, when the normalised pressure drop first reaches unity the average coordination number $Z_m \sim 3.05$ and when, in the case of the wall bounded system, the pressure drop overshoot reaches a maximum value $Z_m \sim 3.15$. As explained in Sect. 2.2.1, in 2D the critical coordination number $Z_c = 3$ for $\mu = \infty$ but for finite friction (in this case $\mu = 0.3$) it depends on the percentage of sliding contacts and is slightly higher.

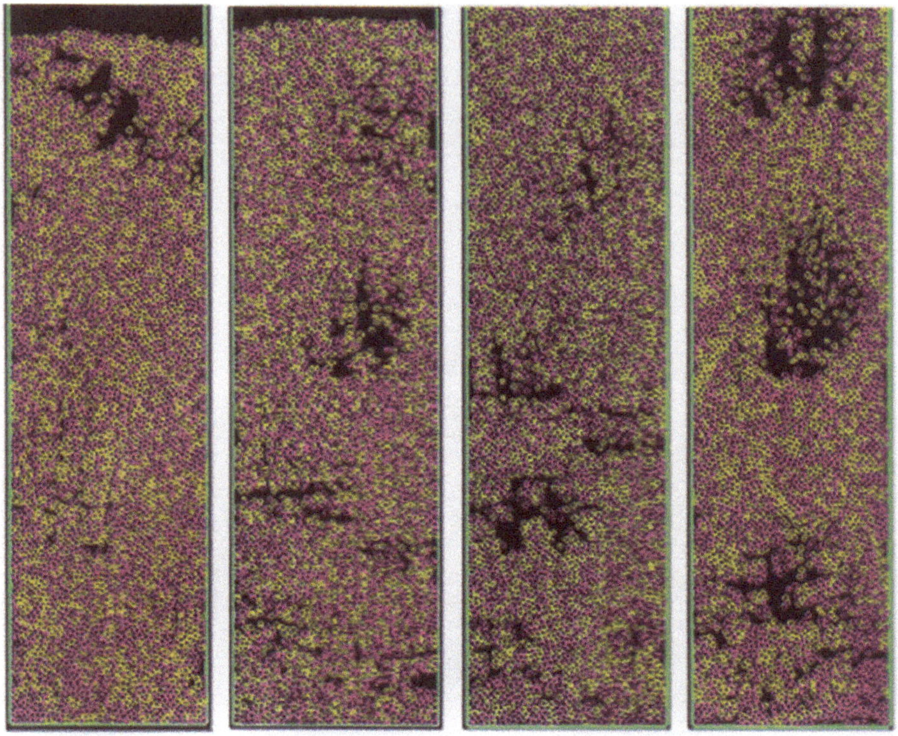

Fig. 7.13 Particle configurations at U_{mb}

7.3 3D Simulations

This section presents results of 3D DEM-CFD simulations of a bubbling fluidised bed the motivation for which was to examine bubble formation, bubble rise velocity and bubble splitting, see Kafui et al. (2006) for more details. The bed consists of 100,000 elastic-plastic spherical particles in the size range 50 ± 5 μm which were randomly generated in a container of 2×2 mm cross-section and pluvially deposited to yield a powder bed of height 2.9 mm and void fraction $\varepsilon = 0.404$. The bed was then fluidised using a superficial gas velocity of 14.3 mm/s (~4.8 U_{mf}). Figure 7.16 illustrates the initial bed expansion, showing particles and contacts.

In order to clearly visualise bubble behaviour, the bed was divided into eight thin vertical slices, each about $5d_p$ thick, and videos were produced for each slice to examine the evolution of the particle configuration, fluid velocities, particle velocities and interparticle contacts. Snapshots obtained from the videos are used below to illustrate the behaviour.

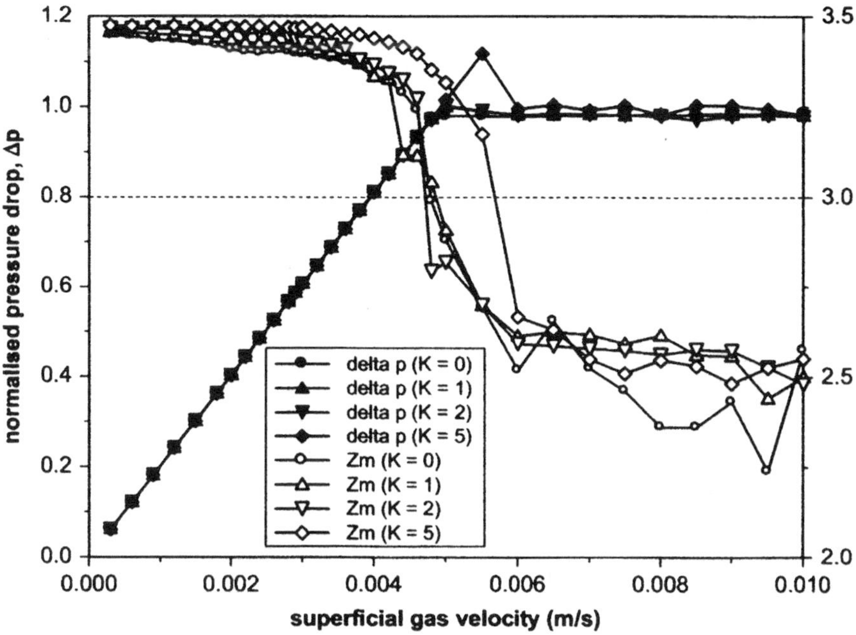

Fig. 7.14 Evolution of pressure drop Δp and mechanical coordination number Z_m

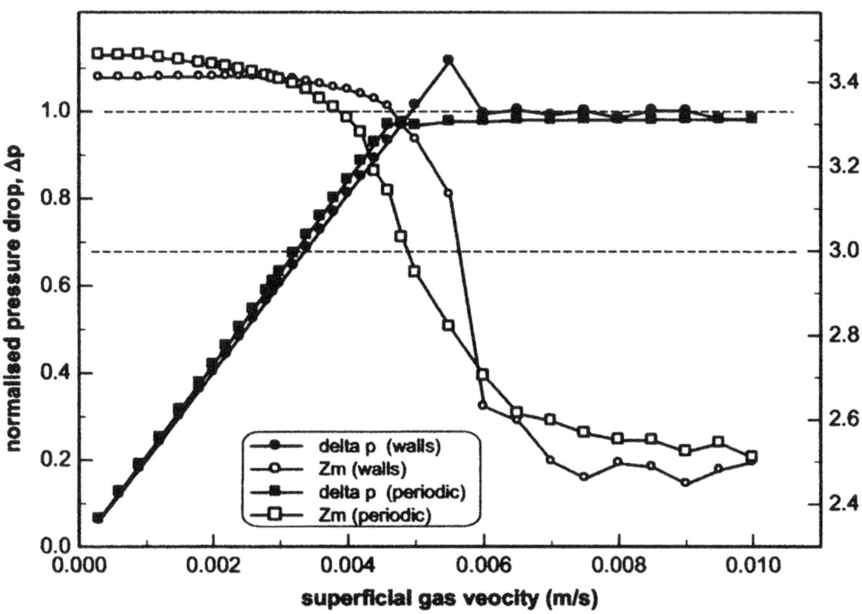

Fig. 7.15 Effect of boundary constraint

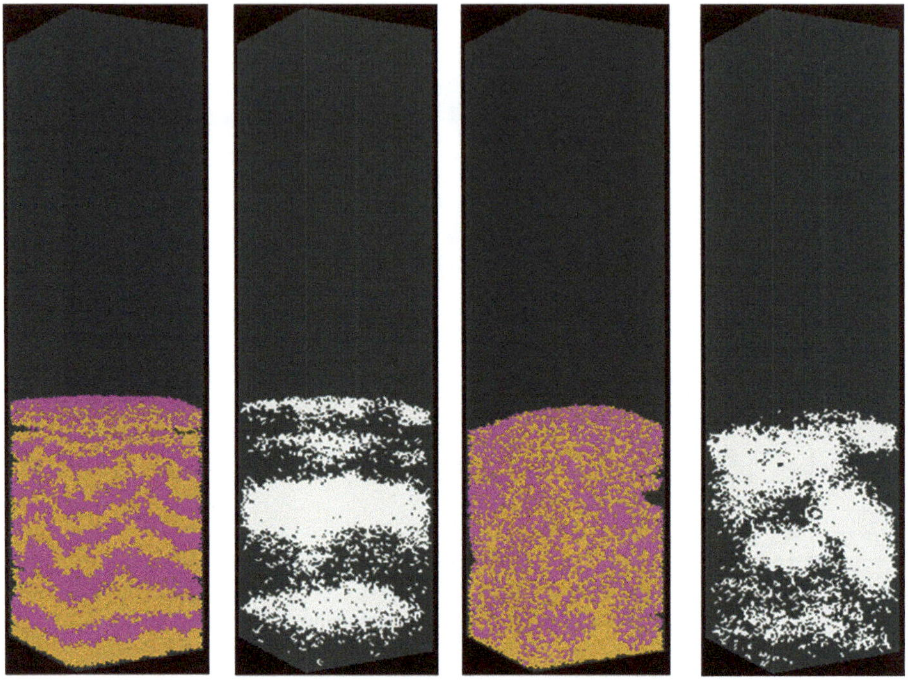

Fig. 7.16 Initial bed expansion (showing particles and contacts)

7.3.1 Bubble Formation

Figure 7.17 shows the typical formation of a bubble. The figure reveals that the bubble was initially seeded from horizontal air pockets spanning the width of the bed and originating from the base. In general, when the particle flow is mainly downwards along the walls, the air pockets are swept upwards with the upward-flowing central core towards the axis of the bed and coalesce to form a single well-defined bubble which either erupts at the top of the bed or occasionally splits into two bubbles. The width-spanning air pockets are a start-up phenomenon due to the narrow bed and the idealised uniform fluidising gas inflow used in the simulations. Visual evidence indicates that later in the simulations they are replaced by smaller air pockets.

It can also be seen from Fig. 7.17 that redirection of the fluid flow through the forming bubble becomes more pronounced as the voidage at the core of the bubble increases. From examination of video sequences it was observed that fluid recirculation loops formed at the sides of a fully formed bubble. These recirculation loops are formed by fluid which, on passing through the bubble roof, circulates through the particles adjacent to the bubble and returns again to the bubble. This feature was predicted by Davidson and Harrison (1963) for cases where the bubble rise velocity is greater than the fluidising gas velocity, as in this case.

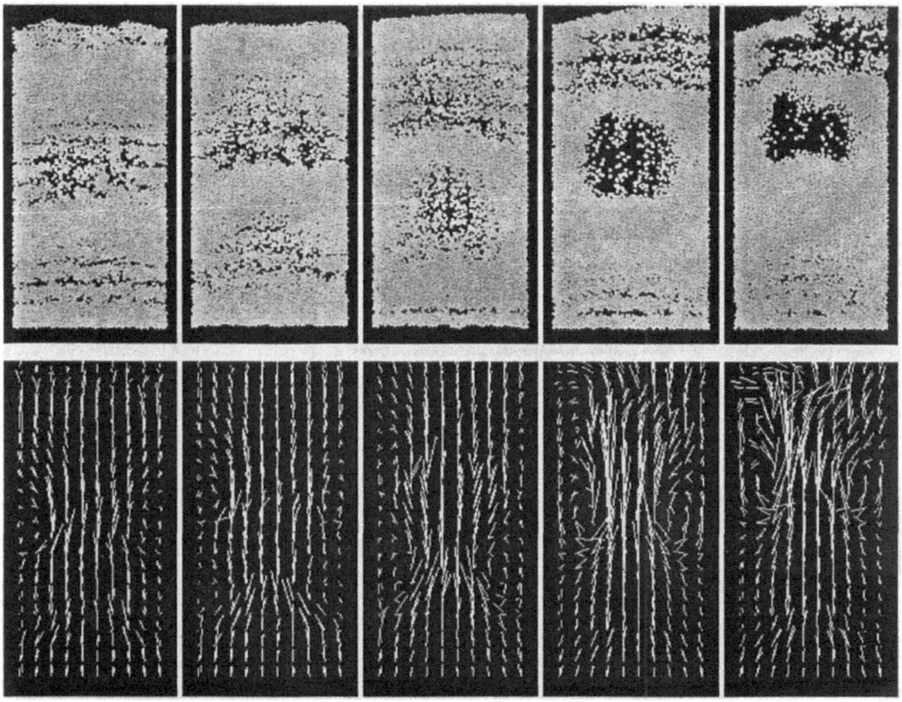

Fig. 7.17 Bubble formation showing particles (*top*) and fluid flow field (*bottom*)

For a well-defined bubble, the particle configuration, fluid velocity field, particle velocity field and location of interparticle contacts, i.e. collisions, are illustrated in Fig. 7.18. It is interesting to note that collisions only tend to occur below the bubble as a consequence of the fluid recirculation loops.

The particle velocity field, as shown in Fig. 7.18, appears to suggest that particles enter the bubbles from the wake of the bubble, as suggested by Rice and Wilhelm (1958). However, as illustrated in Fig. 7.19b, although the absolute particle velocities at the bottom of the bubble are travelling upwards, so is the bubble. Consequently, the bubble velocity was subtracted from the particle velocities to provide the relative particle velocities shown in Fig. 7.19c. It can be seen that the relative velocities indicate that the particles enter the bubble from the roof, as suggested by Harrison et al. (1961). Also it is noted that the relative velocities in the wake of the bubble tend to be horizontal leading to the collisions illustrated in Fig. 7.18.

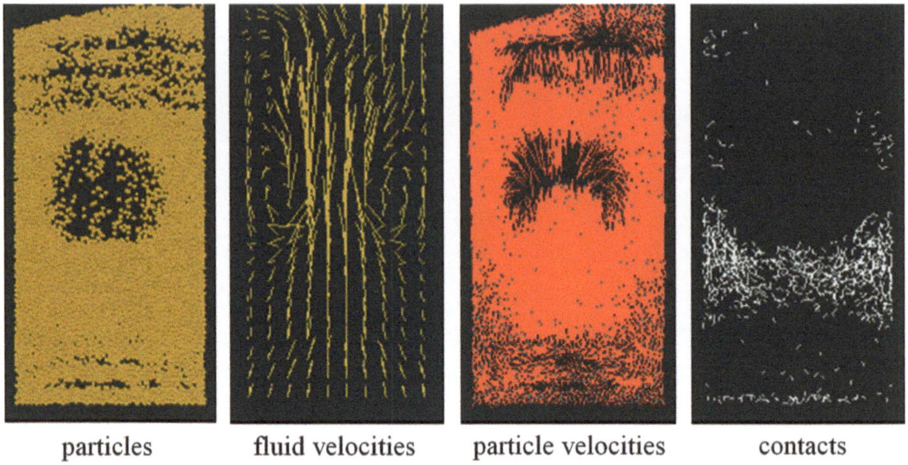

Fig. 7.18 Snapshots showing particles, fluid velocity vectors, particle velocity vectors and interparticle contacts (collisions)

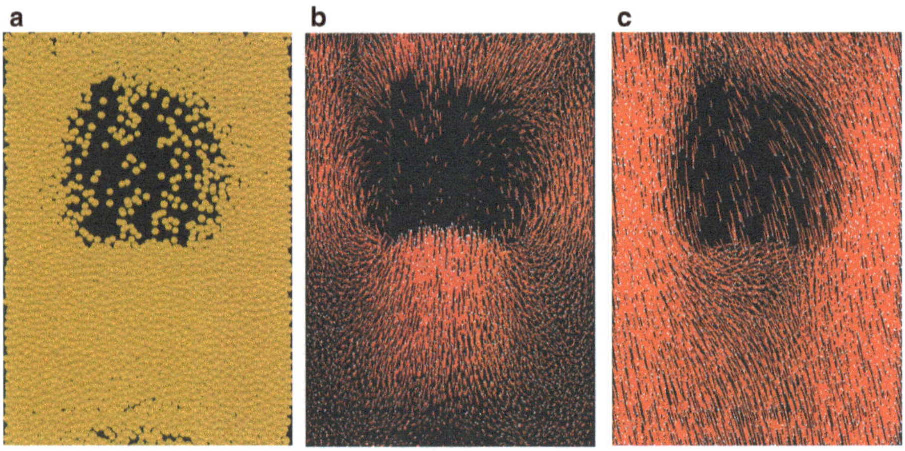

Fig. 7.19 Bubble details showing (**a**) particles (**b**) absolute particle velocities (**c**) particle velocities relative to bubble velocity

7.3.2 Bubble Rise

Bubble dimensions (x_b, y_b, z_b) and bubble velocities (v_{bx}, v_{by}, v_{bz}) in the three coordinate directions can be determined visually for bubble rise sequences such as the example shown in Fig. 7.20. The diameter of a volume-equivalent sphere d_b, which is often employed in defining bubble size, is then calculated from

$$d_b = \frac{x_b + y_b + z_b}{3} \tag{7.19}$$

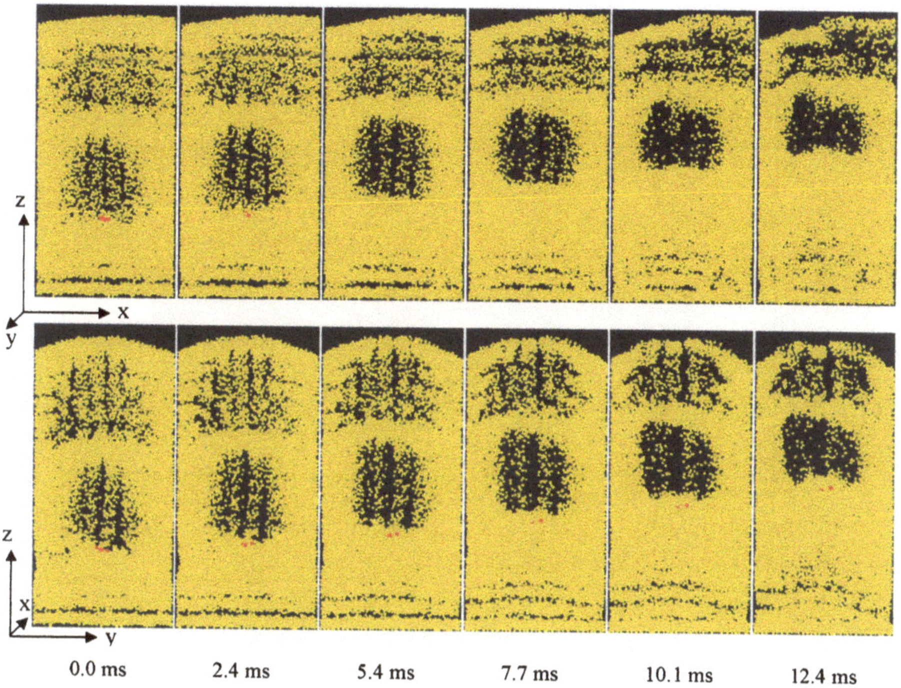

Fig. 7.20 Bubble rise viewed in the y-direction (*top*) in the x-direction (*bottom*)

and the bubble speed v_b is obtained from

$$v_b = \sqrt{v_{bx}^2 + v_{by}^2 + v_{bz}^2} \tag{7.20}$$

Figure 7.20 Bubble rise viewed in the y-direction (top) in the x-direction (bottom)

For the sequence shown in Fig. 7.20, using Eqs. (7.19) and (7.20) the average bubble size was found to be 1.067 mm and the average vertical bubble velocity $v_{bz} = 0.076$ m/s. However, this method of determining bubble size, position and hence velocities is very sensitive to the visual judgement of the bubble boundaries since the shape of the bubble changes as it travels upwards through the bed. An alternative is illustrated in Fig. 7.21.

The linear decrease in pressure from the bottom to the top of a fluidised bed is interrupted by the presence of a bubble, as shown in Fig. 7.21. It can be seen that, within the bubble, the pressure is almost constant. The intersection of the two pressure profiles can be used to identify the location of the bubble centre and the bubble rise velocities as the bubble moves upwards through the bed. The bubble size can also be estimated from the width of the almost zero pressure gradient region of the pressure profile.

7.3.3 Bubble Splitting

As bubbles travel upwards though the bed, bubble splitting and bubble coalescence frequently occur. An example of bubble splitting is illustrated in Fig. 7.22. The

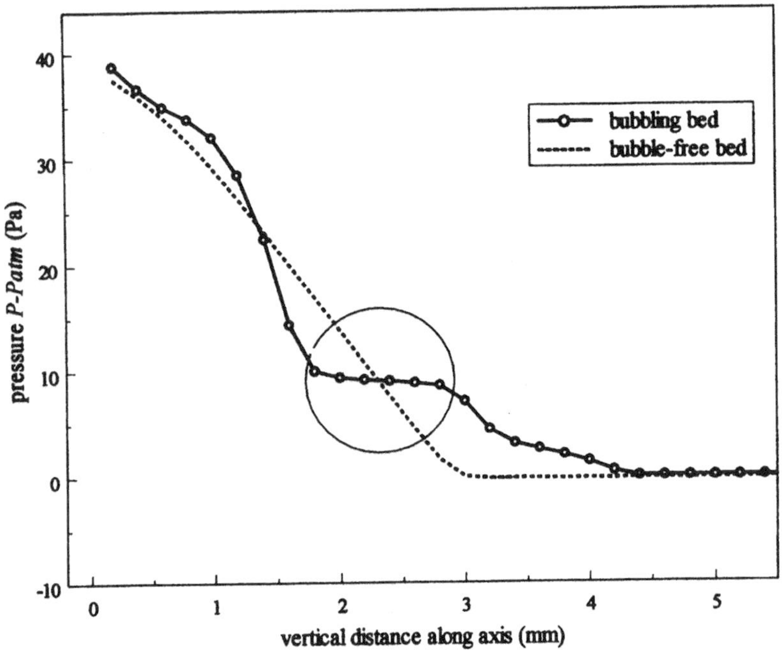

Fig. 7.21 Comparison of pressure profiles through bubbling and bubble-free beds

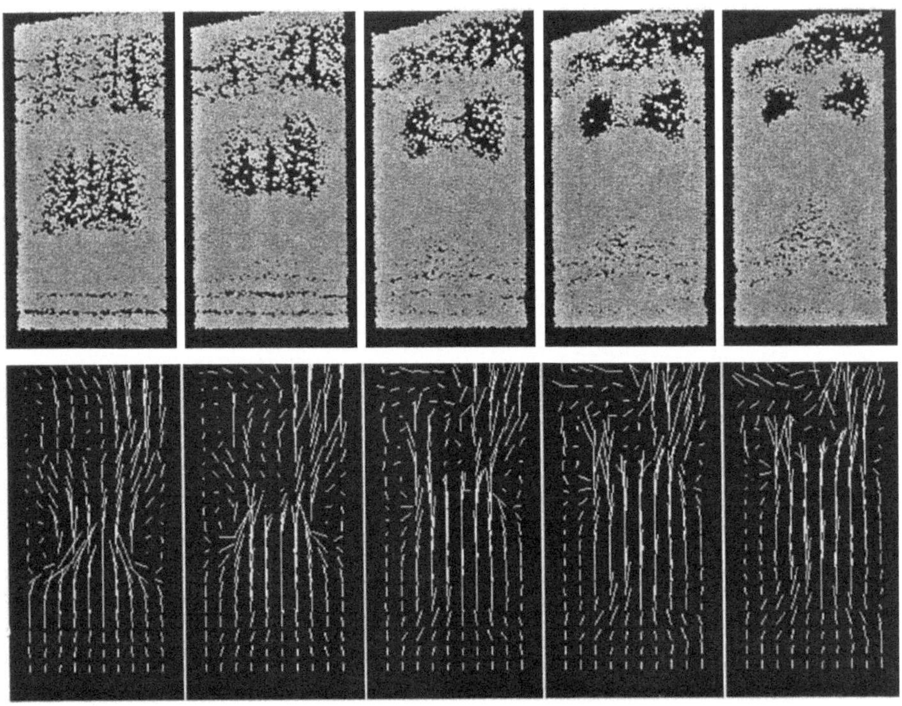

Fig. 7.22 Bubble splitting showing particles (*top*) and fluid velocity vectors (*bottom*)

bubble is fully formed between the first two snapshots in the sequence. However, as can be seen in the second snapshot, particles falling down from the centre of the roof of the bubble initiate a bifurcation of the fluid flow inside the bubble. Consequently, the drag forces acting on the central falling particles is reduced and the concentration of particles raining down from the centre of the roof of the bubble increases, resulting in the creation of two distinct bubbles, as seen in the final snapshot shown in Fig. 7.22. Bubble coalescence was never observed during these simulations and it probably requires a wider and deeper bed to be simulated for this to occur.

References

Anderson, T.B., Jackson, R.: A fluid mechanical description of fluidised beds. Ind. Eng. Chem. Fundam. **6**, 527–539 (1967)

Bi, H.T., Ellis, N., Abba, I.A., Grace, J.R.: A state-of-the-art review of gas-solid turbulent fluidization. Chem. Eng. Sci. **55**, 4789–4825 (2000)

Bird, R.B., Stewart, W.E., Lightfoot, E.N.: Transport Phenomena. Wiley, New York (1960)

Davidson, J.F., Harrison, D.: Fluidised Particles. Cambridge University Press, Cambridge (1963)

Di Felice, R.: The voidage function for fluid-particle interaction systems. Int. J. Multiphase Flow **20**, 153–159 (1994)

Ergun, S.: Fluid flow through packed columns. Chem. Eng. Prog. **48**, 89–94 (1952)

Godard, K., Richardson, J.F.: Bubble velocities and bed expansions in freely bubbling fluidised beds. Chem. Eng. Sci. **24**, 663–670 (1969)

Harrison, D., Davidson, J.F., de Kock, J.W.: On the nature of aggregative and particulate fluidisation. Trans. Inst. Chem. Eng. **39**, 202–211 (1961)

Kafui, K.D., Thornton, C., Adams, M.J.: Discrete particle-continuum fluid modelling of gas-solid fluidised beds. Chem. Eng. Sci. **57**, 2395–2410 (2002)

Kafui, K.D., Thornton, C., Seville, J.P.K.: Numerical simulations of fluidised behaviour of Geldart Group A powder beds. In: Proc. 10th Int. Conf. on Multiphase Flow in Industrial Plant, Tropea, Italy, pp. 571–577 (2006)

Rice, W.J., Wilhelm, R.H.: Surface dynamics of fluidised beds and quality of fluidisation. AICHE J. **4**, 423–429 (1958)

Thornton, C., Yang, F., Seville, J.P.K.: A DEM investigation of transitional behaviour in gas-fluidized beds. Powder Technol. **270**, 128–134 (2015)

Tsuji, Y., Kawaguchi, T., Tanaka, T.: Discrete particle simulation of two-dimensional fluidized bed. Powder Technol. **77**, 79–87 (1993)

Yang, F., Thornton, C., Seville, J.P.K.: Effect of surface energy on the transition from fixed to bubbling gas-fluidised beds. Chem. Eng. Sci. **90**, 119–129 (2013)

Yerushalmi, J., Cankurt, N.T.: Further studies of the regimes of fluidization. Powder Technol. **24**, 187–205 (1979)

Yerushalmi, J., Cankurt, N.T., Geldart, D., Liss, B.: Flow regimes in vertical gas-solid contact systems. AIChE Symp. Ser. **174**, 1–12 (1978)

Chapter 8
Quasi-static Deformation

Abstract This final chapter deals with quasi-static deformation of compact particle systems with enduring contacts. Although DEM can be applied to more dynamic problems, as illustrated in previous chapters, the original BALL and TRUBAL codes were intended to be used to examine quasi-static deformation in the context of soil mechanics. Prior to the introduction of DEM, analyses were restricted to regular arrays of equal-sizeed spheres and an example is provided at the start of this chapter. This is followed by presentations of the results of 2D simulations of the direct shear, simple shear and biaxial compression tests in order to examine and illustrate shear localisation, shear bands and non-coaxiality of stress, fabric and strain rate during simple shear deformation. Results of 3D periodic cell simulations of axisymmetric compression are then presented to illustrate the evolution of the induced structural anisotropy and coordination number, the significance of the strong force sub-network and the effect of plastic deformation at the interparticle contacts. Finally, general 3D states of stress are explored and, for radial deviatoric straining, stress and fabric response envelopes are illustrated and a deviatoric flow rule is identified. This final section of the chapter provides an excellent example of the power of DEM simulations, due to the ability to simulate an infinite number of different tests from exactly the same initial state, a problem that is impossible to achieve by an experimentalist.

Prior to the introduction of DEM simulations by Cundall and Strack (1979), particle based analyses of quasi-static deformation was restricted to regular arrays of equal-sized spheres. In the 1950s Mindlin's research group used their contact mechanics solutions, see Sect. 3.1, to obtain differential stress-strain relationships for face-centred cubic arrays (Duffy and Mindlin 1957), simple cubic arrays (Deresiewicz 1958) and close-packed hexagonal arrays (Duffy 1959). An alternative approach was to consider regular arrays of rigid spheres in order to rationalise shear strength in terms of the micromechanics at the particle scale. This approach was initiated by Rennie (1959) who considered the conditions under which sliding occurred at all contacts under axisymmetric compression. This problem was also examined by Rowe (1962) and, in the context of plane strain, by Leussink and Wittke (1963). A more comprehensive analysis of failure conditions for body-centred orthorhombic

C. Thornton, *Granular Dynamics, Contact Mechanics and Particle System Simulations*, Particle Technology Series 24, DOI 10.1007/978-3-319-18711-2_8

arrays of rigid spheres subjected to general 3D straining was provided by Thornton and Barnes (1982), which will be presented in the following section.

8.1 Failure Conditions for Regular Arrays of Rigid Spheres

Consider a range of regular arrays of rigid spheres that can be classified as body-centred orthorhombic; and include body-centred tetragonal, body-centred cubic and face-centred cubic arrangements as special cases. The structure of the body-centred orthorhombic array is shown in Fig. 8.1a. The central sphere is in contact with eight other spheres whose centres are defined by the coordinates

$$X_i = l_i D \quad (i=1,2,3) \tag{8.1}$$

where D is the diameter of the spheres and l_i are the direction cosines of the branch vector joining the centres of each of these spheres with the centre of the central sphere, which is located at the origin of the Cartesian coordinate system shown. Consider irrotational deformation for which the principal stress and strain-rate tensors are coaxial. Coaxiality is associated with a "multiple sliding" mechanism in which sliding occurs at all contacts and the eight spheres remain in contact with the central sphere throughout the subsequent deformation. Furthermore, for irrotational deformation, the magnitude of the relative displacement-rate, $\dot{\Delta}$, is the same at all contacts and so we need only consider the relative displacement of one of the spheres, as shown in Fig. 8.1b. Hence, if the assembly is subjected to a uniform strain-rate $\dot{\varepsilon}_{ij}$, the relative displacement-rate of the contacting sphere may be expressed as

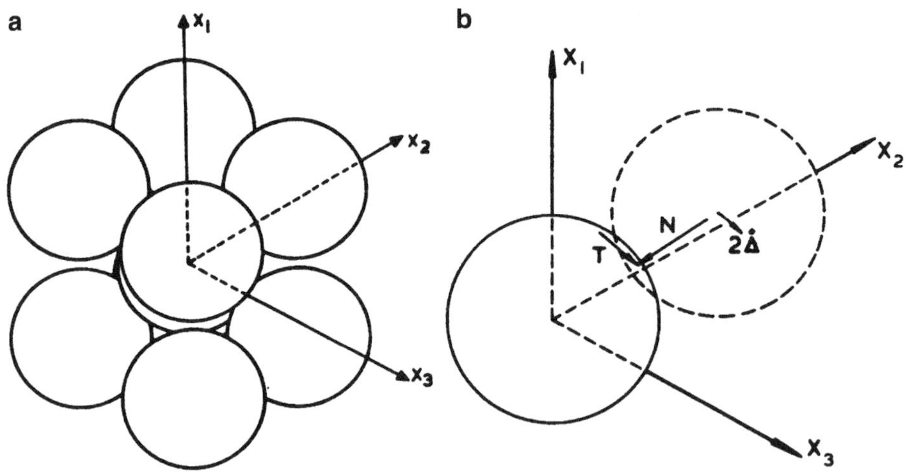

Fig. 8.1 Body-centred orthorhombic array (**a**) configuration (**b**) contact details

$$\dot{X}_i = \dot{\varepsilon}_{ij} X_j \tag{8.2}$$

or

$$\dot{X}_i = m_i 2\dot{\Delta} \tag{8.3}$$

where m_i are the direction cosines of the displacement-rate vector. Using Eqs. (8.1), (8.2) and (8.3) we obtain the strain-rate tensor

$$\dot{\varepsilon}_{ij} = \frac{m_i 2\dot{\Delta}}{l_j D} \tag{8.4}$$

in which $\dot{\varepsilon}_{ij} = 0$ for $i \neq j$, i.e. the principal strain-rate directions coincide with the axes of the Cartesian reference frame.

Since the tractions acting on the surface of a typical sphere are discrete forces acting at the points of contact with adjacent spheres, the state of stress within a sphere is not homogeneous. However, due to the regular arrangement of the spheres, the average stress tensor will be identical for all spheres and will be equivalent to the macroscopic state of stress for a large assembly of spheres. The macroscopic stress tensor is defined by

$$\sigma_{ij} = \frac{1}{V}\int_V \sigma_{ij} dV = \frac{1}{V}\sum_{1}^{n} x_i P_j \tag{8.5}$$

where V is the volume of the space occupied by each sphere on which there are n discrete forces P acting at the contacts defined by the coordinates x_i. It is convenient to partition the stress tensor as follows

$$\sigma_{ij} = \alpha_{ij} + S_{ij} \tag{8.6}$$

where α is the normal force contribution and S is the tangential force contribution to the stress tensor. For irrotational deformation there is no particle spin and hence, in order to satisfy moment equilibrium for each sphere, all the tangential forces must be equal. It then follows that, since coaxiality is associated with sliding at all contacts, all the normal forces must also be equal. Therefore, let N be the normal force and $T = \mu N$ be the tangential force at each contact.

If the direction cosines of the contact normal vector are n_i ($= l_i$ for spheres) then the normal and tangential force contributions are obtained as

$$\alpha_{ij} = \frac{1}{V}\sum_{1}^{n} x_i N_j = \frac{4ND}{V} n_i n_j \tag{8.7}$$

and

$$S_{ij} = \frac{1}{V}\sum_{1}^{n} x_i T_j = \frac{4\mu ND}{V} n_i m_j \tag{8.8}$$

in which $\alpha_{ij} = S_{ij} = 0$ for $i \neq j$. Substituting Eqs. (8.7) and (8.8) into (8.5)

$$\sigma_{ij} = \sigma_{kk}\left(n_i n_j + \mu n_i m_j\right) \text{ with } \sigma_{kk} = \frac{4ND}{V} \tag{8.9}$$

Considering the rate of energy dissipation per unit volume, we may write

$$\sigma_{ij}\dot{\varepsilon}_{ij} = \left(\alpha_{ij} + S_{ij}\right)\dot{\varepsilon}_{ij} \tag{8.10}$$

Substituting Eqs. (8.4), (8.7) and (8.8) leads to $\alpha_{ij}\dot{\varepsilon}_{ij} = 0$ and $S_{ij}\dot{\varepsilon}_{ij} = 8T\dot{\Delta}/V$ demonstrating that the normal force contribution is the non-dissipative stress and the tangential force contribution is the dissipative stress.

The stress tensor, defined by Eq. (8.9), applies to all body-centred orthorhombic arrays and defines the states of stress that will cause yield, the yield mechanism being defined by the strain-rate tensor, given by Eq. (8.4). Thus, having specified the physical properties of the array (structure and interparticle friction) it is possible, for a given mean stress, to identify the complete range of stress states that will cause yield by varying the direction of the relative displacement-rate vector $\dot{\Delta}$ at the contacts.

8.1.1 Solutions

Yield conditions are conventionally represented in principal stress space by yield surfaces. Due to the dependence on mean stress it is clear that the yield surfaces are cones and it is then convenient to depict the yield conditions on the deviatoric plane (σ_{kk} = constant) of principal stress space, as illustrated in Fig. 8.2. For the body-centred cubic array $\left(l_1 = l_2 = l_3 = 1/\sqrt{3}\right)$ the yield conditions are defined by a circle centred on the space diagonal, as shown in Fig. 8.2a. The radius of the circle is equal to $\mu\sigma_{kk}/\sqrt{3}$ and, according to Eq. (8.4), all the states of stress defined by the circular yield locus are associated with zero rate of volumetric strain. The yield conditions for body-centred tetragonal arrays (e.g. $l_1 \neq l_2 = l_3$) are defined by ellipses, as shown in Fig. 8.2b, which are symmetric about one of the axes and whose centres are displaced from the space diagonal. Figure 8.2c shows a typical yield locus for a body-centred orthorhombic array $(l_1 \neq l_2 \neq l_3)$.

Although there are an infinite number of possible body-centred orthorhombic arrays there are restrictions due to the particle shape in that, for spheres, the directions cosines are restricted to the range $1/2 \leq l_i \leq 1/\sqrt{2}$. Consequently

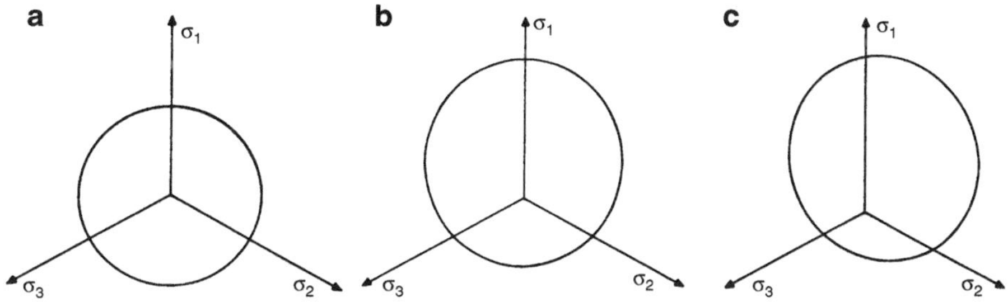

Fig. 8.2 Deviatoric yield loci (**a**) body-centred cubic (**b**) body-centred tetragonal (**c**) body-centred orthorhombic

Fig. 8.3 Limit envelope

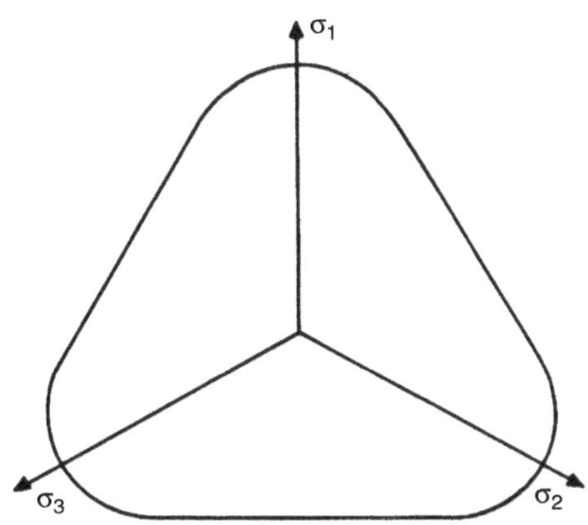

there is a limit to the possible locations of the yield loci on the deviatoric plane. Figure 8.3 shows the limiting envelope that circumscribes all the limiting yield loci. It is interesting to note that the envelope is rather similar to the experimentally observed failure envelope of Lade and Duncan (1975), see also Sect. 8.3.2.

The yield locus shown in Fig. 8.2c is redrawn in Fig. 8.4a and a three dimensional view is shown in Fig. 8.4b. The yield conditions for all body-centred orthorhombic arrays (with the exception of the body-centred cubic case) are defined by oblique elliptical cones with their base planes parallel to the deviatoric planes. Figure 8.4b illustrates how the total stress vector is subdivided into its non-dissipative, α, and dissipative, S, components. It can be seen that the non-dissipative stress vector coincides with the axis of the cone and the dissipative stress vector lies in the deviatoric plane. The corresponding components on the deviatoric plane are shown in Fig. 8.4a which illustrates that only part of the deviatoric stress is dissipative since

$$s_{ij} = a_{ij} + S_{ij} \tag{8.11}$$

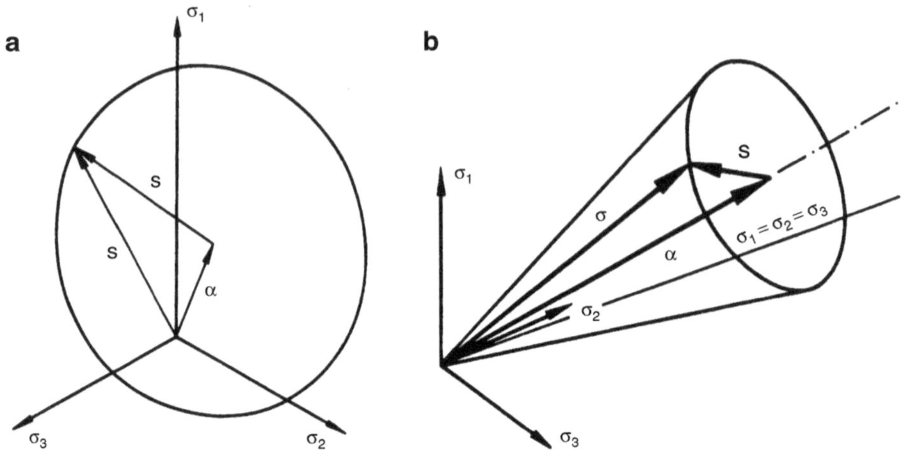

Fig. 8.4 Geometry of a typical yield surface

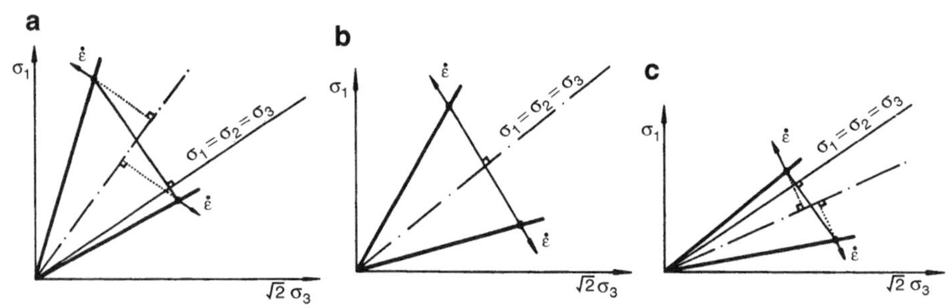

Fig. 8.5 Yield cones for body-centred tetragonal arrays

where the deviatoric stress

$$s_{ij} = \sigma_{ij} - \sigma_{kk}\delta_{ij}/3 \tag{8.12}$$

and

$$a_{ij} = \alpha_{ij} - \alpha_{kk}\delta_{ij}/3 \tag{8.13}$$

Yield cones for three body-centred tetragonal arrays $(l_2 = l_3)$ are shown in Fig. 8.5 with the strain-rate vectors for axisymmetric stress states $(\sigma_{22} = \sigma_{33})$ superimposed. It is clear that the normality rule of classical plasticity theory does not apply. Instead, the strain-rate vectors are normal to the axes of the cones. This is a physical requirement that the strain-rate and non-dissipative stress vectors are orthogonal.

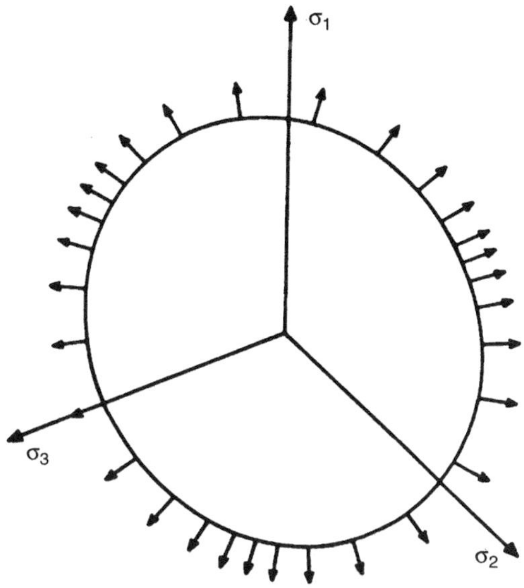

Fig. 8.6 Dissipative stress plane viewed along the axis of the cone

It follows from the geometry in Fig. 8.5 that planes orthogonal to the axis of the cone may be referred to as "dissipative stress planes. Adopting this terminology, it is found that normality of the strain-rate vector to the yield surface is only observed when the dissipative stress planes are examined by looking down the axes of the cones. This is illustrated in Fig. 8.6, for the case of the body-centred orthorhombic solution shown in Fig. 8.2c.

As shown in Fig. 8.2, the location and shape of the deviatoric yield loci depend on the structure of the array. Yield results in a change in structure but, for irrotational deformation, the general body-centred orthorhombic description remains valid throughout the strain softening deformation process. Consequently it can be inferred from Figs. 8.2, 8.4, and 8.5 that, during deformation, the non-dissipative stress vector (which coincides with the yield cone axis) rotates about the origin of principal stress space and, hence, the deviatoric yield loci translate, distort and rotate in a manner that reflects the changes in the structural anisotropy of the array.

Figure 8.7 illustrates the evolution of the deviatoric yield locus for a face-centred cubic array that is deformed in an arbitrary manner into a body-centred cubic arrangement.

The initial yield locus is centred at A with the stress state defined by point 1. During deformation the centre of the yield locus follow the trajectory ABCDE with the corresponding stress path defined by the points 12345. The lines A1....E5 correspond to the current dissipative stress vectors and it can be seen that they are tangential to the trajectory ABCDE of the centre of the yield locus. Therefore, the motion of the deviatoric yield locus may be defined as follows

$$\dot{a}_{ij} = -hS_{ij} \tag{8.14}$$

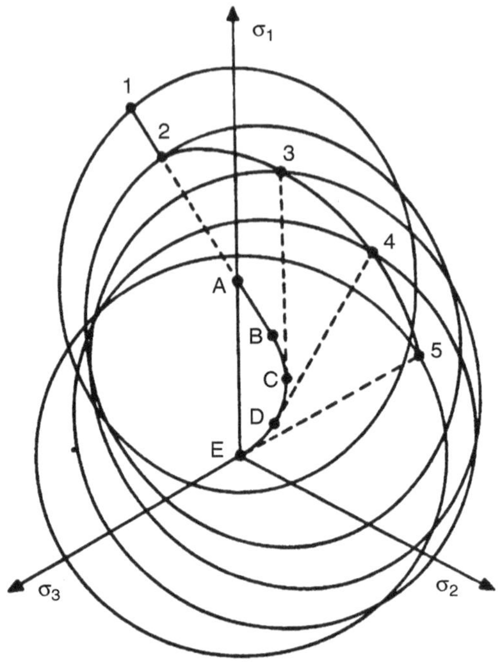

Fig. 8.7 Evolution of the deviatoric yield locus during deformation

which, it is interesting to note, corresponds to Ziegler's (1959) modification to Prager's (1955) kinematic hardening rule.

From the geometrical solutions presented above it is possible to identify the complete corresponding macroscopic constitutive plasticity model in terms of the yield function, flow rule and the (negative) hardening law, see Thornton and Barnes (1982) for details. However, for random arrays of spheres, even monodisperse systems, the above type of analysis is not possible due to the irregular arrangement. This, together with other complications such as particle rotation, particle shape and polydispersity, is what makes real granular materials, even spherical particle systems, so complex and fascinatingly challenging, as will be demonstrated in the remainder of this chapter.

8.2 2D Simulations

Early applications of DEM to quasi-static deformation were restricted to two-dimensional systems (Cundall and Strack 1979; Thornton and Barnes 1986; Rothenburg and Bathurst 1989). Two-dimensional simulations of quasi-static deformation are, in general, of limited use due to the extreme kinematic constraint that there cannot be any out-of-plane motion of the particles. Even in plane strain

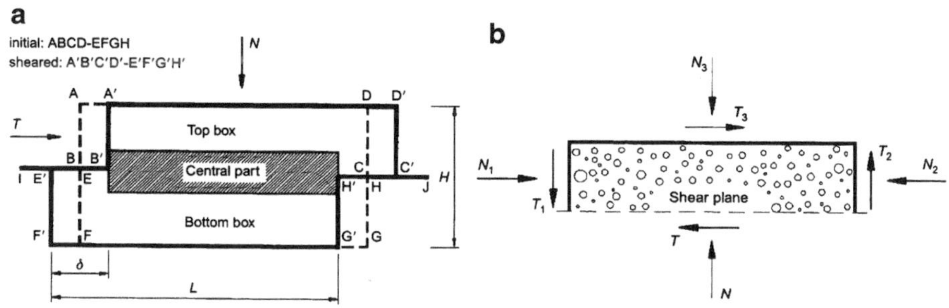

Fig. 8.8 The direct shear test (**a**) schematic diagram (**b**) forces acting on the top half of the specimen

particles are free to move out-of-plane. It simply requires that the average out-of-plane motion is zero.

However, since visualisations of 3D particle systems are extremely difficult to clearly illustrate, 2D simulations can be useful to obtain a basic understanding of certain micromechanical features of granular media. In this sense, the next section considers 2D simulations in the context of strain localisation and shear bands. This is followed by results obtained from 3D simulations in axisymmetric compression, the so-called 'standard triaxial test', followed by an examination of the shear behaviour of compact particle systems under general 3D loading in which $\sigma_1 \neq \sigma_2 \neq \sigma_3$.

8.2.1 Direct Shear Tests

In the direct shear test, the top half of the specimen is translated relative to the bottom half of the specimen in order to create a shear band/plane across the mid-height of the specimen. Although the direct shear test has been criticised for many years, it is still widely used in industry for testing cohesionless material. Traditionally, in process engineering the test is performed in a Jenike shear cell, which is circular in cross-section, whereas the Casagrande shear box, which has a square cross-section, is used in geotechnical engineering. In both cases the externally applied vertical and horizontal forces are measured and the ratio of horizontal to vertical load is assumed to provide an estimation of the average ratio of shear to normal stress acting in the shear band and, thereby, provide a direct measure of the angle of internal friction, although the exact state of stress within the shear band is unknown.

The test arrangement is shown diagrammatically in Fig. 8.8a. The upper half of the specimen is contained by four walls (AB, CD, AD and BI), which are all translated horizontally in the opposite direction to the four walls (EF, GH, FG and CJ) containing the lower half of the specimen. The top and bottom walls are free to move vertically, but not to rotate. In simulations, walls do not interact with other walls, only with adjacent particles. More details can be found in Thornton and Zhang (2003), Zhang and Thornton (2007).

Two-dimensional simulations were performed on a polydisperse system of 5000 elastic spheres. Particle-particle and particle-wall interactions were modelled by the Hertz, Mindlin and Deresiewicz models described in Sect. 3.1. During all simulations no out-of-plane motion was permitted and, for the purpose of calculating volumes, necessary for determining the stress tensor and the porosity, the dimension in the third orthogonal direction was taken to be the average particles diameter, in this case 0.06 mm.

By summing all the forces at the particle/wall contacts the resolved reaction forces may be calculated. Figure 8.8b shows the wall reaction forces for the upper part of the shear box. From equilibrium it follows that

$$T = N_1 - N_2 + T_3 \tag{8.15}$$

$$N = T_1 - T_2 + N_3 \tag{8.16}$$

and the Coulomb definition of the mobilised angle of shear resistance ϕ_c is given by

$$tan\ \phi_c = \frac{\tau}{\sigma_n} = \frac{T}{N} \tag{8.17}$$

The shaded central band shown in Fig. 8.8a, which is about ten mean particle diameters wide, is an approximation to the anticipated shear zone that will develop at the mid-height of the specimen. Consequently, an approximation to the state of stress acting in the shear zone can be obtained from the distribution of contact forces in the shaded region using Eq. (2.45).

Sample preparation is simulated by omitting the top wall, AD in Fig. 8.8a, and randomly generating the particles as a granular gas (no contacts) in a rectangle whose height is approximately 40 % higher than the box height, AF in Fig. 8.8a. By introducing gravity the particles rain down to form a bed. The top wall is then positioned above all the particles and moved downwards to establish multiple ball/wall contacts. The gravity field is then removed.

In order to simulate constant vertical normal stress tests, the top and bottom walls are permitted to move in the vertical direction. Initially, tests were simulated by adjusting the vertical position of the top and bottom walls to keep the vertical normal force N_3 constant, see Fig. 8.8b. However this did not work, because the vertical tangential forces T_1 and T_2 cannot be controlled by moving the top wall. Ideally one would wish to maintain constant the vertical normal stress acting in the shear band but, for reasons given later, this is not possible. Consequently the stress tensor for the complete specimen was calculated using Eq. (2.45) and the top and bottom walls were continuously adjusted to maintain the average vertical normal stress σ_v, for the complete specimen, at a constant value. To achieve this, at each time step, the vertical velocity of the top wall was set to

$$\dot{u}_v = g\left(\sigma_v^d - \sigma_v^c\right) \tag{8.18}$$

where σ_v^d is the desired vertical normal stress, σ_v^c is the calculated vertical normal stress for the whole specimen and g is the gain parameter, in this case set to 0.001.

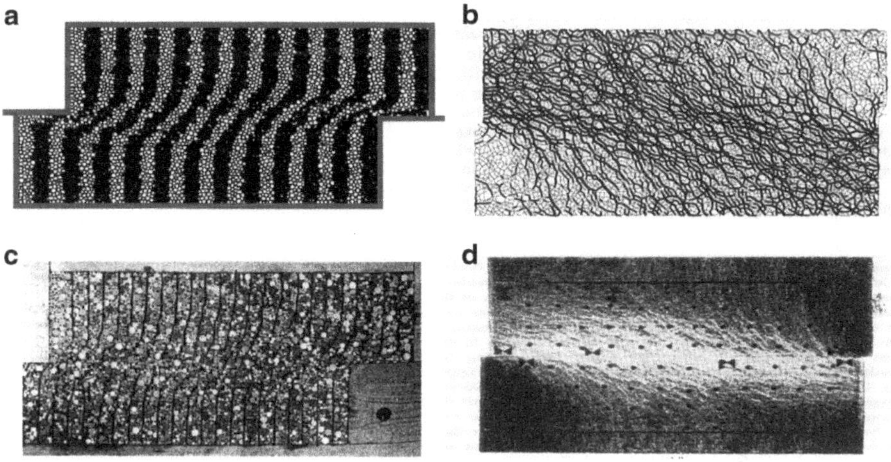

Fig. 8.9 Deformation pattern (**a**) and force transmission pattern (**b**) compared with experiments by (**c**) Vardoulakis and Sulem (1995) and (**d**) Dyer and Milligan (1984)

The vertical velocity of the bottom wall is set to minus the value calculated by Eq. (8.18). If $\dot{u}_v$ is greater than a specified maximum admissible velocity $\dot{u}_{max}$ then

$$\dot{u}_v = sign\left[\dot{u}_{max}, g\left(\sigma_v^d - \sigma_v^c\right)\right] \tag{8.19}$$

with $\dot{u}_{max}$ set to 8 μm/s.

Figure 8.9a shows the particle configuration at the end of shearing, which may be compared with a real experimental result shown in Fig. 8.9c. By colour-banding the particles vertically before the start of shear it can be seen that deformation is localised in a narrow shear zone located at the mid-height of the specimen. The actual shape of the shear zone is lenticular, not rectangular as indicated diagrammatically in Fig. 8.8a. The width of the shear band in the centre is wider than near the edge. It is clear that, within the lenticular shear zone, the shear strain is greatest near the edges and reduces towards the centre. Consequently the state of stress is not homogeneous within the shear zone.

The force transmission through the system of particles, at peak stress ratio, is illustrated in Fig. 8.9b. Each line is drawn between the centres of two particles in contact with each other. The larger-than-average contact forces are indicated by black lines and grey lines indicate the lower-than-average contact forces. The magnitude of each force is indicated by the thickness of the line, scaled to the current maximum contact force. It can be seen from the figure that there are significantly large contact forces acting on the top left-hand wall and the bottom right-hand wall. Only small forces are transmitted to the other two vertical walls. On the top and bottom walls the large forces concentrate to the side near the end walls that push the sample. At the other side of the top and bottom walls the forces are relatively small. It is clear that the normal stress distribution along the top and

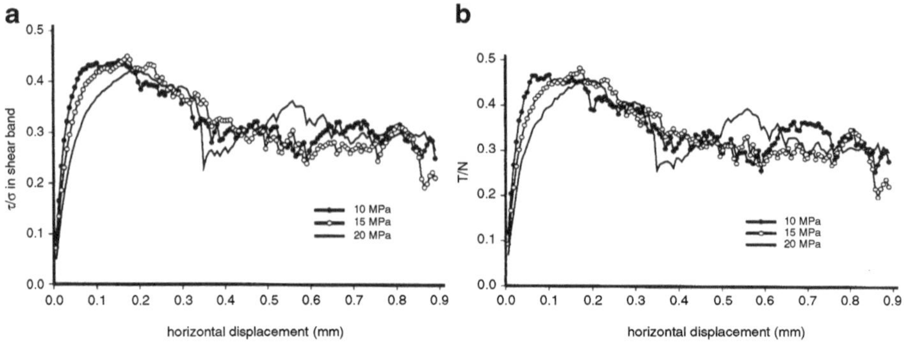

Fig. 8.10 Mobilised shear strength as defined (**a**) by the average stress ratio in the central part and (**b**) by the force ratio T/N

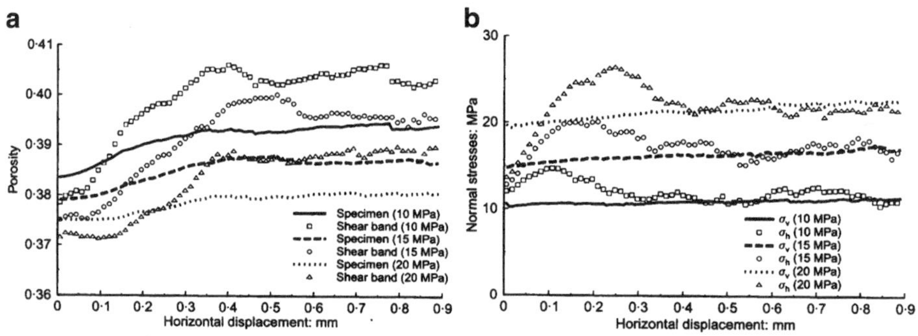

Fig. 8.11 Evolution of (**a**) porosity and (**b**) vertical and horizontal normal stresses in the shear band

bottom walls is not uniform. Consequently, the strong force transmission pathways are inclined as they pass through the shear zone at the mid-height of the specimen. The overall pattern clearly indicates a very heterogeneous distribution of stress that correlates very well with photoelastic observations for crushed glass, shown in Fig. 8.9d, in which the orientation of the light stripes approximate to the major principal stress direction.

The evolution of the mobilised shear strength obtained from three direct shear simulations in which the normal stress was maintained constant at 10, 15 and 20 MPa is shown in Fig. 8.10. It can be seen that, in spite of the gross inhomogeneity within the specimen, the evolution of the mobilised shear strength defined by T/N is similar to that obtained from calculating the average stress ratio in the central part of the specimen.

Figure 8.11a shows the evolution of the porosity of the complete specimen, as determined from the vertical displacements of the top and bottom walls, and the porosity changes that occurred in the central region used to approximate the shear band. Although the calculated porosity changes in the shear band show significant

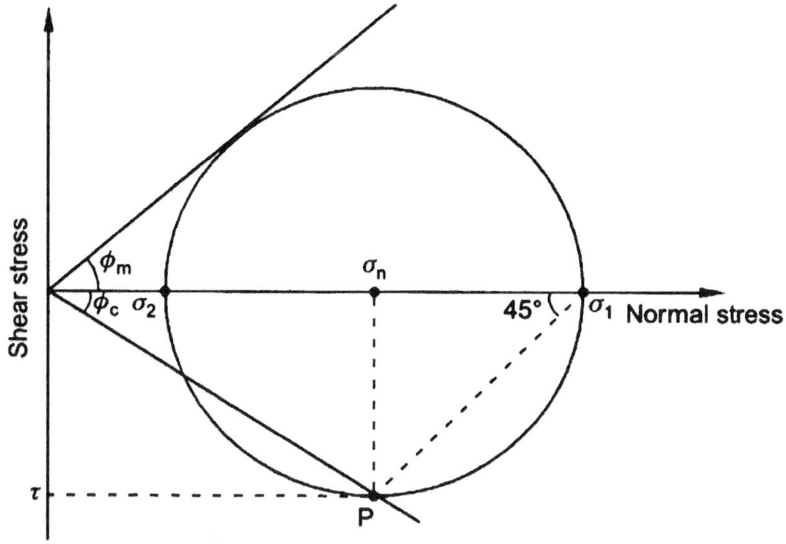

Fig. 8.12 Steady state stress conditions in the shear band

fluctuations, it is clear that the rate of dilation and the final voidage in the shear band exceeds that indicated by the wall movements.

Figure 8.11b shows the evolution of the vertical and horizontal normal stresses acting in the central part of the specimen used to approximate the shear band. Although the average vertical normal stress of the complete specimen was controlled to remain constant it can be seen that the vertical normal stress in the approximated shear band increases by about 10 % during the simulations. As the relative displacement between the upper and lower parts of the shear box increases the horizontal normal stress increases until the stress ratio τ/σ_n is a maximum. At this stage the horizontal normal stress is greater than the vertical normal stress reflecting the fact that the strong force chains, shown in Fig. 8.9b, are inclined at an angle less than 45° to the horizontal. After peak stress ratio the horizontal normal stress decreases and, significantly, becomes approximately equal to the vertical normal stress when the steady state is attained at the end of the tests. If the vertical and horizontal normal stresses in the approximated shear band are equal then σ_n defines the centre of the Mohr circle at the steady state, the radius of which is equal to τ and the stress combination (τ,σ_n) is given by point P in Fig. 8.12. Note that the Mohr definition of the mobilised angle of shearing resistance $\phi_m \neq \phi_c$ since

$$sin\, \phi_m = \frac{\sigma_1 - \sigma_2}{\sigma_1 + \sigma_2} \tag{8.20}$$

and, at the steady (critical) state, $tan\, \phi_c = sin\, \phi_m$.

In the standard shear testing procedures using the Jenike shear cell, the sample is sheared in two steps. In the first, the sample is 'presheared' under a constant normal

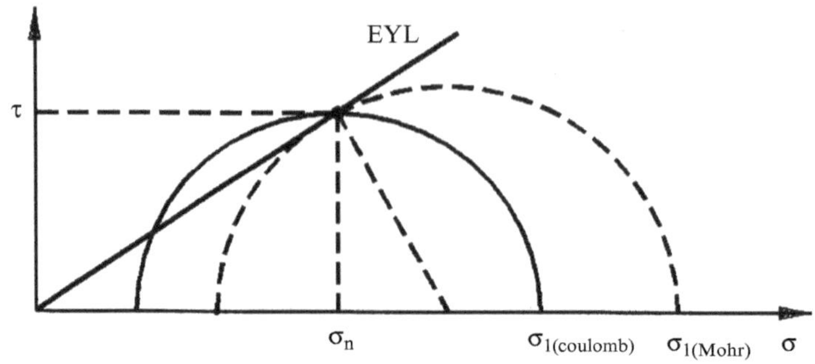

Fig. 8.13 Possible Mohr circles of stress at the steady state

stress σ_n until steady state flow is established and τ remains constant. It is normal practice to assume that the measured values of σ_n and τ define a point on the 'effective yield locus' (EYL) shown in Fig. 8.13.

From the experimental data the location of the corresponding Mohr stress circle is not known. It is conventionally assumed that the 'effective yield locus' is the same as the Mohr-Coulomb line traditionally used in soil mechanics, implying that $\phi_c = \phi_m$ contrary to the inequality illustrated in Fig. 8.12. Consequently, the EYL is assumed to be tangential to the Mohr circle as illustrated by the larger circle shown in Fig. 8.13 with the 'major consolidation stress' $\sigma_1 = \sigma_{1(Mohr)}$. However, the simulation data shown in Fig. 8.11b indicate that the actual Mohr circle is the smaller of the two circles and the 'major consolidation stress' is $\sigma_1 = \sigma_{1(Coulomb)}$. From the geometry of Fig. 8.13, it follows that

$$\frac{\sigma_{1(Mohr)}}{\sigma_{1(Coulomb)}} = (1 + \sin\phi_c)\frac{(1 + \tan^2\phi_c)}{(1 + \tan\phi_c)} \tag{8.21}$$

which indicates that the Mohr interpretation over predicts the major principal stress and that, as a consequence, the corresponding 'flow function' under predicts the 'unconfined yield stress' for a given value of major principal stress.

8.2.2 *Shear Bands*

In this section, biaxial compression tests are used to examine strain localisation and shear band formation, using polydisperse systems of 5000 elastic spheres with interparticle friction $\mu = 0.5$ and either kinematically controlled wall boundaries or periodic boundaries.

Figure 8.14 shows a wall-bounded system with a porosity of 0.406. The initial configuration, prior to shearing, is shown in Fig. 8.14a in which the particles are colour banded to enable subsequent identification of possible shear bands. Figure 8.14b shows the system with zero wall friction at 30 % deviator strain. Note that the direction of compression is horizontal. It is evident that there is an inclined shear band running from the top right-hand corner to a position along the left-hand side

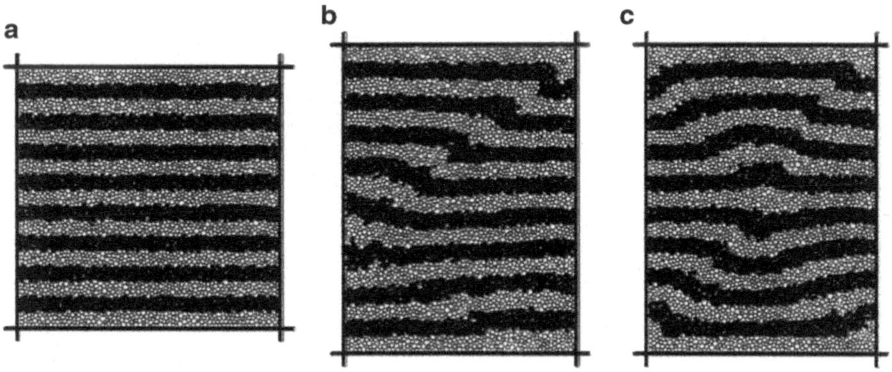

Fig. 8.14 Wall bounded specimens (a) before shear (b) end of shear ($\mu_w = 0.0$) and (c) end of shear ($\mu_w = 0.5$)

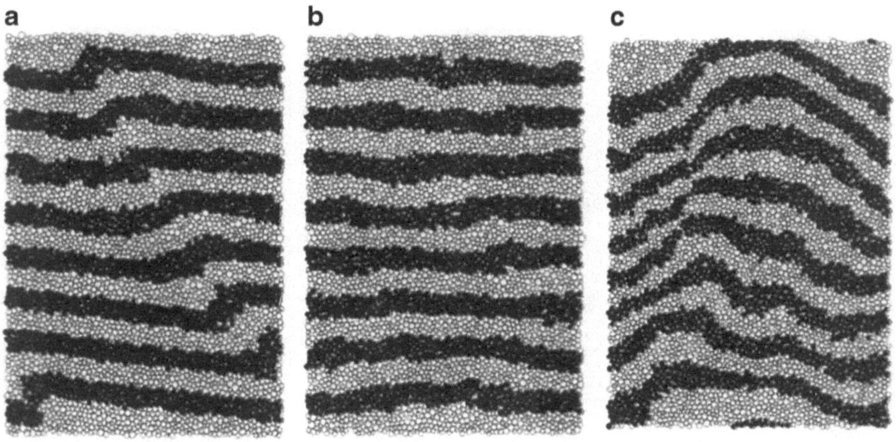

Fig. 8.15 Periodic cell specimens with initial porosities (a) 0.365 (b) 0.401 (c) 0.436

wall. When the wall friction coefficient was changed to $\mu_w = 0.5$, two conjugate intersecting bands are observed, Fig. 8.14c, emanating from near the corners of the specimen. The figure demonstrates that the inclination of shear bands depends on the boundary conditions.

Results of periodic cell simulations are illustrated in Fig. 8.15 for three different initial porosities. A distinct shear band developed in the densest specimen, as shown in Fig. 8.15a. Figure 8.15b is typical of medium dense specimens with no evidence of shear band formation. For the loosest system, Fig. 8.15c suggests a folding mechanism similar to that observed in rock formations. Interestingly, this system never established a percolating pattern of force transmission across the system with enduring contacts. Throughout the test, most particle interactions were collisional corresponding to fluid-like behaviour.

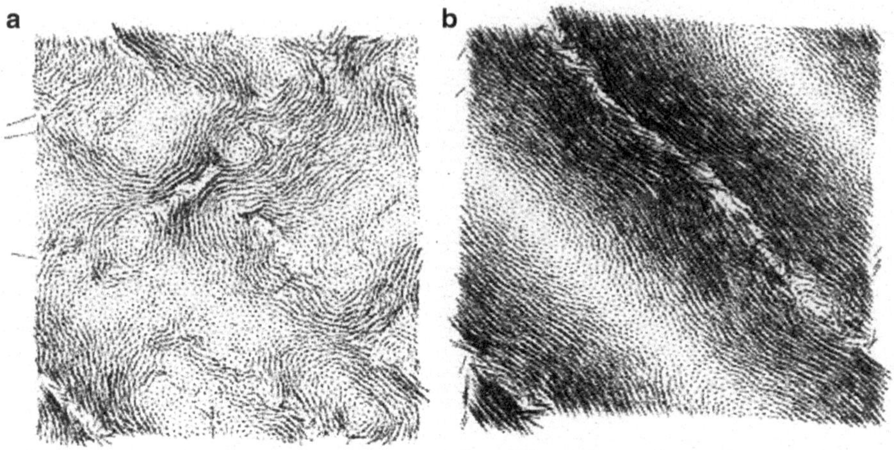

Fig. 8.16 Incremental displacement fields (**a**) pre-peak (**b**) post-peak

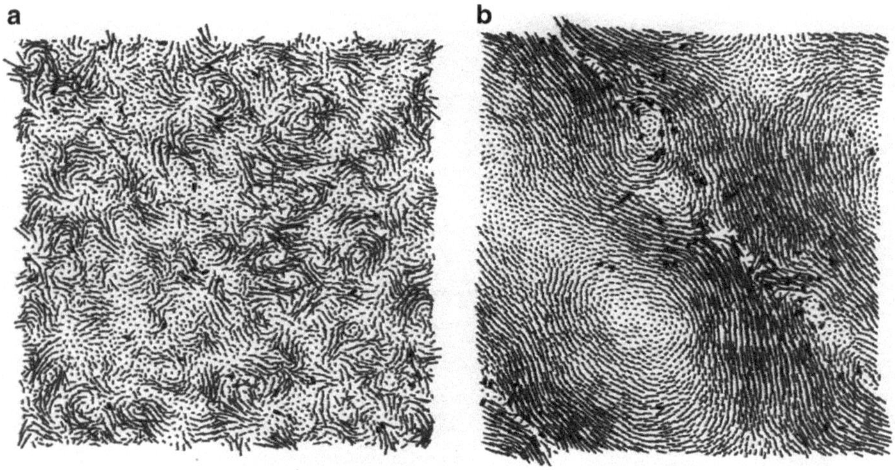

Fig. 8.17 Velocity fluctuations (**a**) near the start of shear (**b**) just after peak

Further detailed information can be obtained about shear bands, as illustrated in Figs. 8.16, 8.17, and 8.18 for the densest specimen illustrated in Fig. 8.15a. Particle locations at any two loading steps can be used to calculate displacement increments. In Fig. 8.16a the incremental displacement field is shown for an increment in deviator strain of about 2 % just prior to attaining peak stress ratio. As expected, no clear shear band is observed at this pre-peak stage. The incremental displacement field is essentially uniform except for a few hints of localization taking place randomly inside the system. Figure 8.16b shows that during post-peak deformation there are large incremental displacements distributed along an inclined band, which corresponds to the finally developed shear band shown in Fig. 8.15a.

Fig. 8.18 Particle rotation field (**a**) pre-peak (**b**) post-peak

Figure 8.17 shows the fluctuating velocity field near the start of shear and just after peak stress ratio. The velocity fluctuations are defined as $\widetilde{v}_i = v_i - \langle v_i \rangle$, where v_i is the particle velocity and $\langle v_i \rangle$ is the average particle velocity. In a periodic cell, if the origin is taken to be the centre of the cell, the average particle velocity is zero and the fluctuating velocity is that resulting from the particle contact forces and calculated using Eq. (2.3). Figure 8.17a shows the fluctuating velocity field near the start of shear. It can be seen that the velocity vectors form many, randomly distributed, local circular cells, as first observed by Williams and Rege (1997). As shearing continues the velocity vectors rearrange, the vortex structures enlarge and, after peak stress ratio, the large velocity fluctuations align in opposite directions along a distinct shear zone, as shown in Fig. 8.17b.

Figure 8.18 shows particle rotations during the same time intervals as the displacement increments shown in Fig. 8.16. In the figure, only rotations larger than 10 % of the current maximum particle rotation are shown. Open circles denote clockwise (negative) rotation and solid circles denote anticlockwise (positive) rotation. As shown in Fig. 8.18b, most of the particles in the established shear band exhibit positive rotations. Prior to peak stress ratio, most of the particles exhibiting significant negative rotations align in the direction perpendicular to that of the positive rotations.

It is clear from Figs. 8.16a and 8.18a that the locations of the large particle rotations correlate with the pattern of the incremental displacement field and indicate the existence of small micro-shear bands during strain hardening prior to reaching the maximum deviator stress, as first demonstrated by Kuhn (1999). Thornton and Zhang (2006) also demonstrated that the locations of the intermittent small micro-shear bands prior to peak also correlated with local regions of high dilation.

It should be noted that, in Figs. 8.16, 8.17, and 8.18, there appears to be a second shear band during post-peak deformation, which is located at the bottom left-hand corner. However, this is not correct because of the periodic nature of the

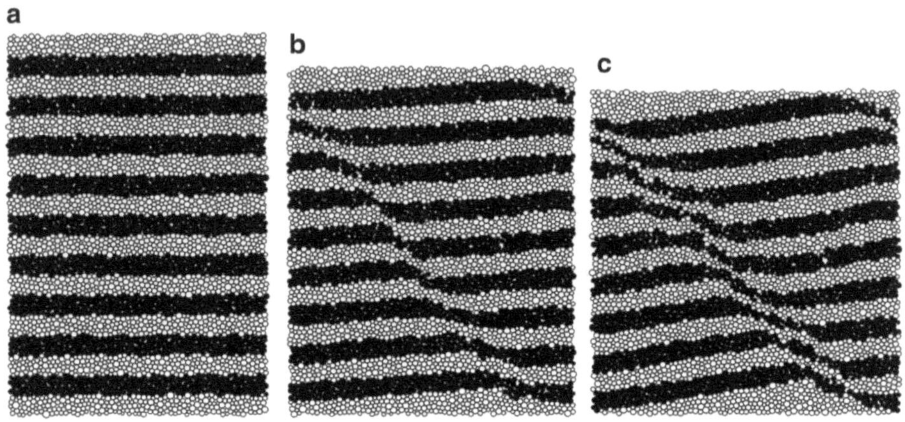

Fig. 8.19 Particle configuration at shear strains of (**a**) 0 % (**b**) 17.6 % (**c**) 33 %

boundaries. This 'short' shear band also exists at the bottom left-hand corners of the cell above and the cell to the right, and is therefore simply a continuation of the one continuous shear band.

Figure 8.19 illustrates the occurrence of a shear band in an initially rectangular periodic cell, Fig. 8.19a, with an aspect ratio of 1.5. The system was compressed vertically, expanded horizontally, and it can be seen in Fig. 8.19b that a shear band formed post-peak inclined at an angle >45° to the major principal plane. At the end of the simulation, however, the shear band inclination had reduced to about 45°, Fig. 8.19c. In a periodic cell, in order to maintain continuity with surrounding cells, the shear band inclination necessarily adjusts as the cell dimensions change and, therefore, the inclination of shear bands cannot be examined by periodic cell simulations. However, the existence of shear bands in periodic cell simulations demonstrates that localization and shear band formation are genuine material behaviours rather than artefacts created by boundary imperfections.

In this section it has been demonstrated that, prior to the attainment of peak shear strength, the deformation becomes heterogeneous at the grain scale as a consequence of the development of conjugate sets of micro-shear bands, which are characterised by high rates of dilation and particle spins. At peak shear strength, buckling of the chains of particles transmitting strong contact forces occurs and this creates strong discontinuities in the fluctuating velocity field that leads to the development of a persistent shear band along one of the existing micro-shear band locations. The other micro-shear bands then disappear due to strain localisation within the dominant shear band.

8.2.3 *Simple Shear*

It is difficult to reliably determine the stress-strain relationship within a shear band due to the small number of particles involved. It is, in fact, questionable as to whether or

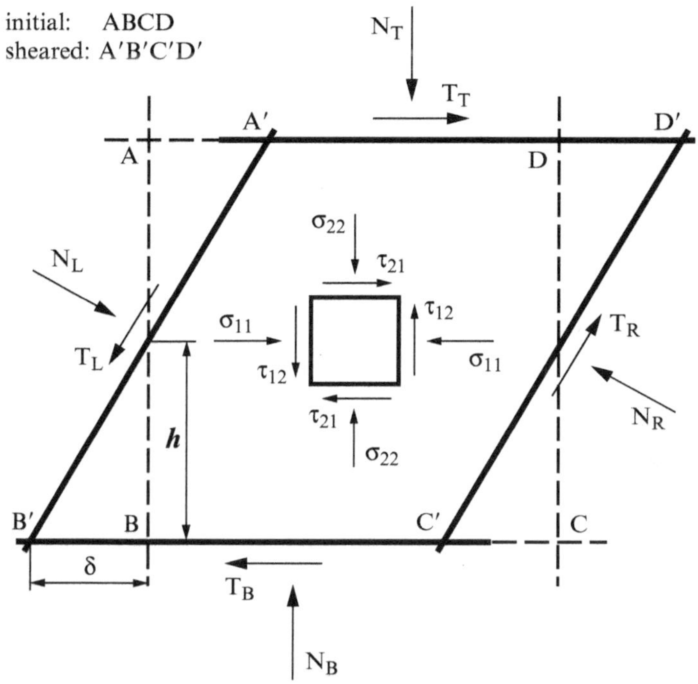

Fig. 8.20 Scheme for simulations of simple shear tests

not a continuum description is appropriate when the width of the band is no more than ten particles. However, it is normally accepted that the mode of deformation within a shear band is one of simple shear. Consequently, simple shear simulations have been performed in order to determine the evolution of the stress tensor, dilation rate, and the principal directions of stress and strain-rate, Thornton and Zhang (2006). This was achieved by simulating systems of particles that were contained within four kinematically controlled walls, analogous to the Grenoble 1γ2ε apparatus, Joer et al. (1992). This ensured overall uniform strain in contrast to the alternative Couette configuration that invariably results in heterogeneous stress and strain states.

The simple shear simulation model is diagrammatically illustrated in Fig. 8.20, in which the dashed lines show the initial positions of the wall boundaries. In order to apply simple shear, the two vertical walls AB and CD are rotated at a constant angular velocity about the mid-points of the two walls. The top and bottom walls AD and BC are translated using a compatible horizontal velocity in opposite directions and, in addition, the vertical velocities are continuously adjusted using servo-control algorithms to ensure that the vertical normal stress component of the stress tensor σ_{22} remains constant. The stress tensor and the fabric tensor are calculated for the system of particles using Eqs. (2.45) and (2.25) respectively, but the summations do not include the contacts with the walls. In the simulations, the wall friction coefficient was the same as the interparticle friction coefficient, i.e. $\mu_w = \mu = 0.5$.

Three specimens were prepared with similar initial porosities but with different initial normal stress ratios, $K_0 = \sigma_h/\sigma_v = 0.5$, 1.0 and 2.0 respectively. Figure 8.21

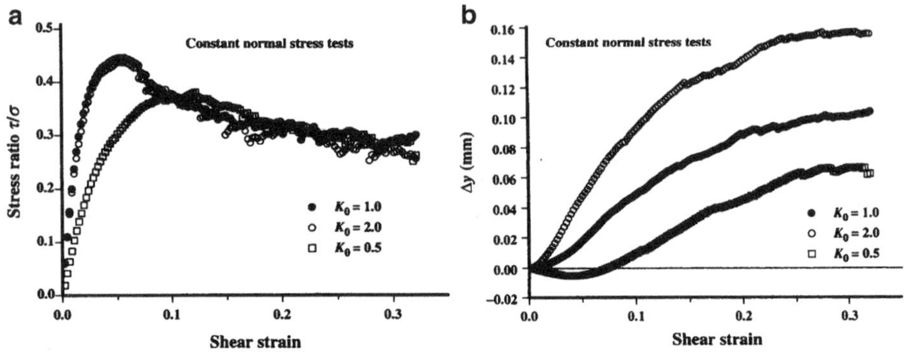

Fig. 8.21 Effect of initial normal stress ratio on the evolution of (**a**) the stress ratio and (**b**) dilation

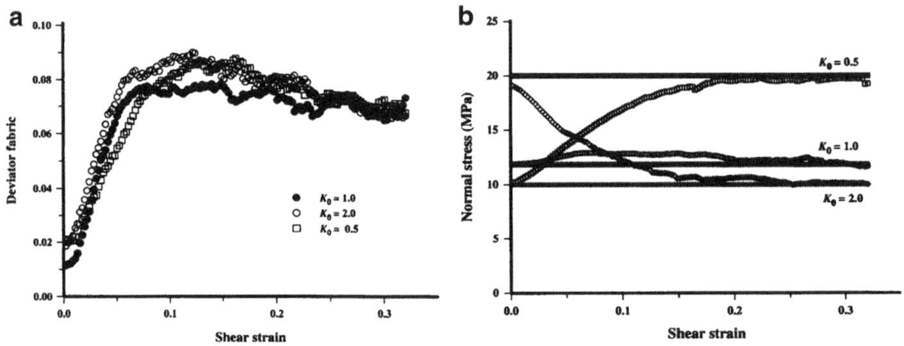

Fig. 8.22 Evolution of (**a**) deviator fabric (**b**) vertical and horizontal normal stresses

shows the evolution with shear strain ($\gamma = \delta/h$) of (a) the ratio of shear to normal stress acting on the horizontal plane τ_{21}/σ_{22} and (b) the corresponding increase in the vertical dimension of the specimen. It can be seen from the figure that the evolutions of the stress ratio for $K_0 \geq 1$ are almost identical but, for the case of $K_0 < 1$, the specimen exhibits a significantly more compliant response and the shear strength is significantly lower. At the end of the tests, the stress ratio is the same for all three specimens and all three are deforming at constant volume. Figure 8.21b shows that, for the $K_0 < 1$ case, there is an initial contraction before the specimen starts to expand. Also, for $K_0 \geq 1$, the specimens dilate from the beginning but at significantly different rates, even though the evolutions of stress ratio are very similar. The results demonstrate that, at least in simple shear, there is no simple ubiquitous relationship between stress ratio and rate of dilation, as usually assumed in traditional soil mechanics literature, Taylor (1948), Rowe (1962) and Bolton (1986). Although there is a significant effect of K_0 on the stress ratio and dilatancy response, Fig. 8.22a demonstrates that the deviator fabric is insensitive to the initial value of K_0.

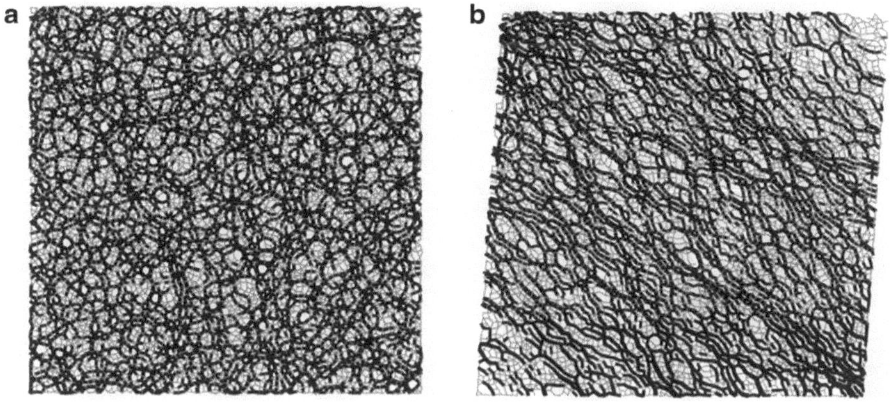

Fig. 8.23 Force transmission (**a**) at start of shear (**b**) at peak shear strength

A major problem with laboratory simple shear experiments normally is the inability to accurately measure the stress σ_h acting parallel to the horizontal, zero-extension direction. This means that the location of the Mohr circle of stress and the principal stress directions are unknown. Figure 8.22b shows the evolution of the two normal stresses, σ_h and σ_v. In the figure the solid symbols indicate the vertical normal stress that is held constant and the horizontal normal stress is indicated by open symbols. The remarkable feature of this figure is that, irrespective of initial stress ratio K_0, the horizontal normal stress increases or decreases in order that the two normal stresses are equal at large strains.

Figure 8.23 illustrates the force transmission at the start of shear and at peak stress ratio for the case $K_0 = 1$. In the figure, the larger than average contact forces are indicated by black lines joining the centres of the two particles transmitting the force and, in a similar way, the less than average contact forces are indicated by grey lines. Figure 8.23a demonstrates that, since the initial state is isotropic, the larger than average forces are randomly orientated. In contrast, Fig. 8.23b shows that the orientation of the larger than average contact forces align themselves with the direction of the major principal stress, indicating that the direction of the major principal stress rotates during simple shear deformation.

A typical Mohr circle representation of the state of stress during simple shear deformation is shown in Fig. 8.24a and the corresponding Mohr circle defining the strain-rate tensor is shown in Fig. 8.24b, where ψ defines the angle of dilation. In Fig. 8.24a the point on the circle with coordinates $(-\tau, \sigma_v)$ defines the shear stress and vertical normal stress acting on the horizontal planes. Drawing a line in the direction of the (horizontal) plane on which this stress combination acts intersects a point on the circle known as the 'pole of the planes', see Schofield and Wroth (1968), p. 295. The property of the pole is that a line connecting any point on the circle with the pole defines the orientation of the plane on which the shear/normal

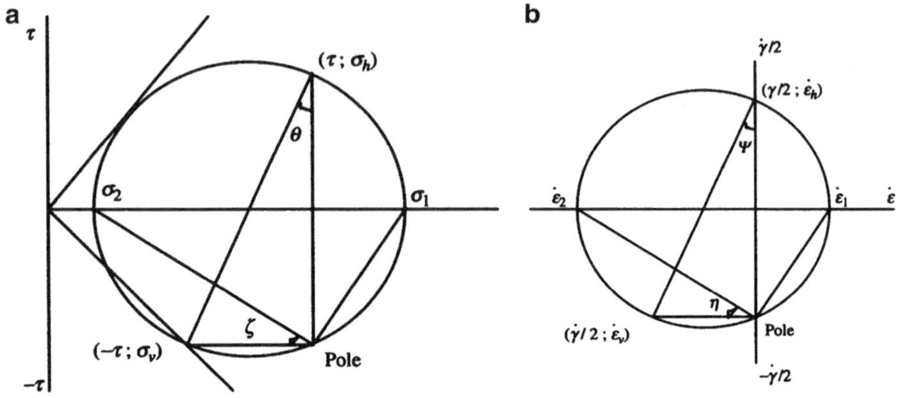

Fig. 8.24 Mohr circle of (**a**) stress (**b**) strain-rate

stress combination represented by that point acts. Figure 8.24a indicates the orientation of the planes on which the two principal stresses act. Consequently, the angle ζ defines the inclination of the major principal stress to the horizontal direction. It follows from Fig. 8.24a that $\zeta = \pi/4 - \theta/2$ where

$$tan\theta = \frac{\sigma_h - \sigma_v}{2\tau} \tag{8.22}$$

Applying the same procedure to Fig. 8.24b we identify that the angle η defining the inclination of the major principal strain-rate to the horizontal direction is $\eta = \pi/4 - \psi/2$ and

$$tan\psi = \frac{\dot{\varepsilon}_h - \dot{\varepsilon}_v}{\dot{\gamma}} = \frac{-\dot{\varepsilon}_v}{\dot{\gamma}} \tag{8.23}$$

It follows from the above that if $\psi = \theta$ then $\eta = \zeta$ and the principal directions of stress and strain-rate are coaxial. Non-coaxiality occurs if $\psi \neq \theta$. The difference between the inclinations of the major principal stress and major principal strain-rate is the angle of non-coaxiality $\iota = \eta - \zeta$. In order to identify ι the procedure is to (i) draw the Mohr circle of strain-rate with the same radius as the Mohr circle of stress, (ii) superimpose the strain-rate circle on the stress circle and (iii) rotate the strain-rate circle so that the points representing the conditions on the horizontal plane $(-\tau, \sigma_v)$ and $(\dot{\gamma}/2,\ \dot{\varepsilon}_v)$ coincide. An example of this for $K_0 > 1$ is illustrated in Fig. 8.25 for (a) expansion ($\psi > 0$) and (b) contraction ($\psi < 0$).

Note that, in simple shear, there is a distinction between the Mohr definition of the angle of shearing resistance given by

$$sin\,\phi_m = \frac{\sigma_1 - \sigma_2}{\sigma_1 + \sigma_2} \tag{8.24}$$

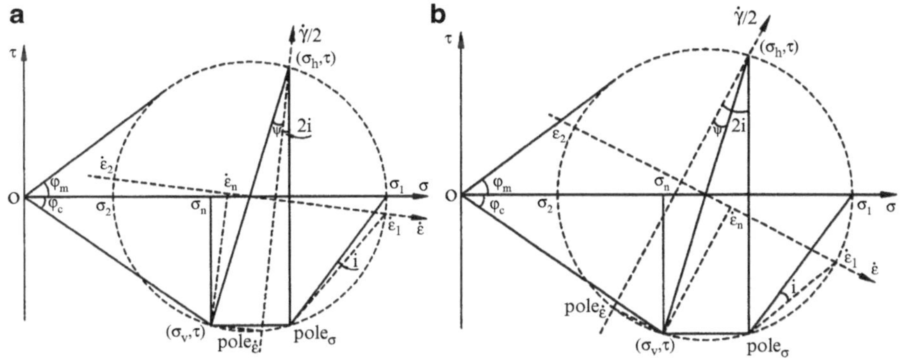

Fig. 8.25 Superimposed Mohr circles of stress and strain-rate (**a**) $\psi > 0$ (**b**) $\psi < 0$

and the Coulomb definition given by

$$tan\,\phi_c = \frac{\tau}{\sigma_v} \tag{8.25}$$

as can be seen in Fig. 8.25. From the geometry of Fig. 8.25a it follows that

$$tan\,\phi_c = \frac{sin\,\phi_m cos\theta}{1 - sin\,\phi_m sin\theta} \tag{8.26}$$

and, from the figure, $\theta = \psi + 2\imath$. Therefore

$$tan\,\phi_c = \frac{sin\,\phi_m cos\,(\psi + 2\imath)}{1 - sin\,\phi_m sin\,(\psi + 2\imath)} \tag{8.27}$$

The above equation relates the stress tensor to the strain-rate tensor and, therefore, can be considered to be the flow rule for simple shear deformation.

For the simple shear simulations with different initial stress states, the evolution of the major principal stress and strain-rate directions (inclination to the horizontal) is shown in Fig. 8.26. The corresponding evolution of the angle of non-coaxiality is shown in Fig. 8.27. From the figures it is noted that (i) if $K_0 \neq 1$ then the initial angle of non-coaxiality is either 0° or 90°, (ii) when the stress ratio is a maximum, cf. Fig. 8.21a, the angle of non-coaxiality $\imath \neq 0$ and (iii) at large strains $\imath = 0$, i.e. the directions of stress and strain-rate are coaxial. Consequently, it is significant to note that at large strains, when the specimens have reached what in soil mechanics terminology is referred to as 'the critical state', coaxiality of stress and strain-rate exists, as illustrated in Fig. 8.28. Remarkably, this phenomenon was first suggested by Hill (1950) but subsequently tended to be ignored by the soil mechanics community.

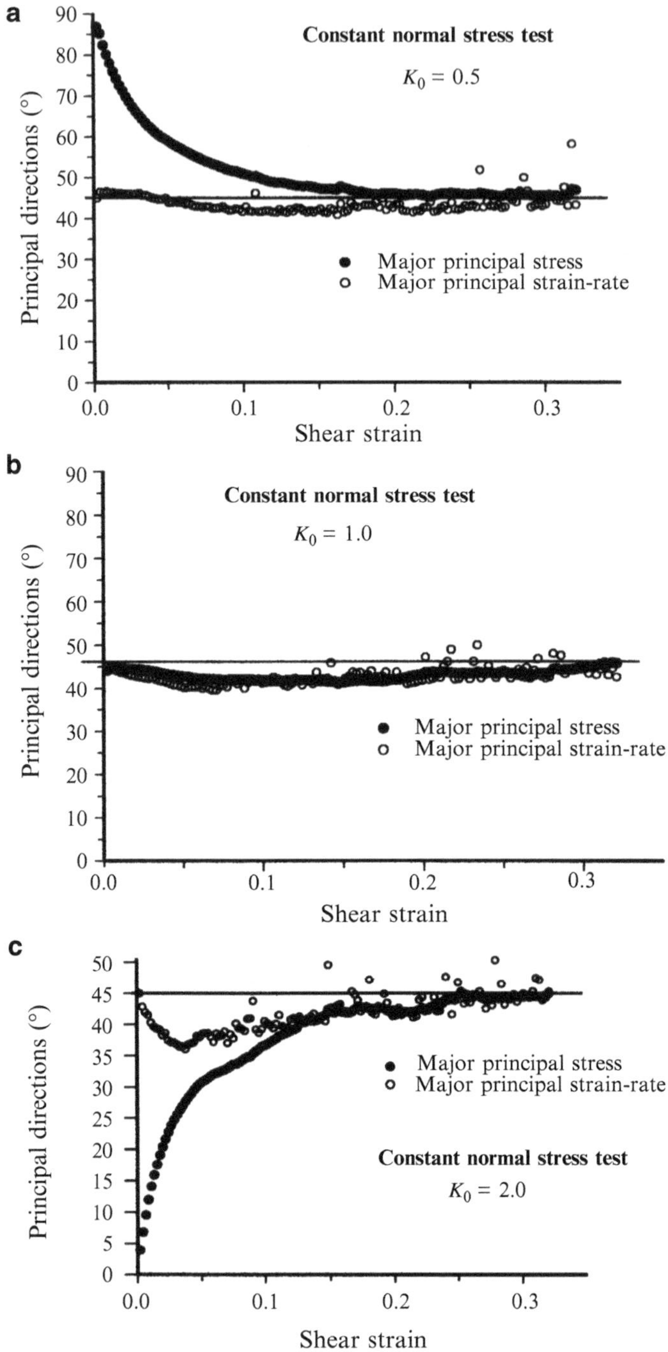

Fig. 8.26 Evolution of the directions of major principal stress and major principal strain-rate

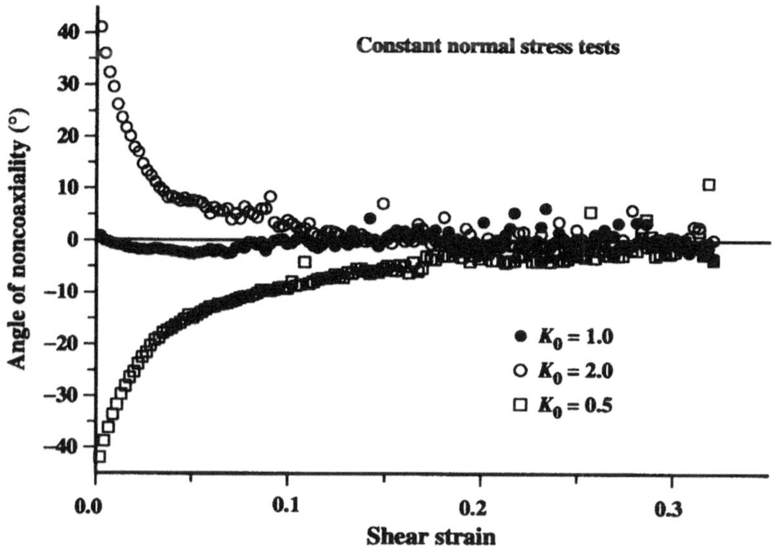

Fig. 8.27 Evolution of the angle of non-coaxiality

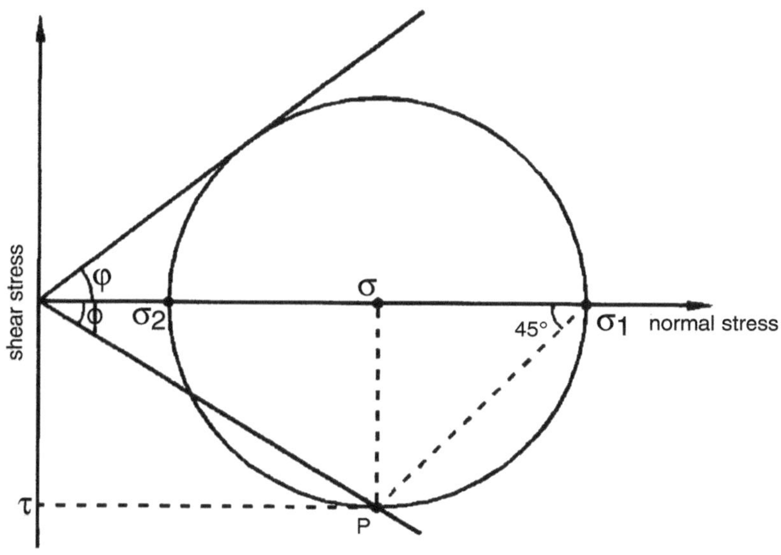

Fig. 8.28 Critical state stress conditions

It has been shown above that, at any stage of shearing, the angle of non-coaxiality depends on the mobilised shear strength, the rate of dilation and the initial stress state. Consequently, the evolution of non-coaxiality is complicated and cannot, in general, be predicted a priori since it is not simply a function of a single material property as suggested by Mandl and Fernandez-Luque (1970).

8.3 3D Simulations

In DEM simulations one may choose to mimic laboratory experiments but, if the objective is to relate the micromechanics to the meso-scale constitutive behaviour, it is necessary to use a periodic cell. In this section all the reported simulations were performed using a representative volume element, with periodic boundaries, subjected to uniform strain fields. In order to control the deformation of the system, a strain-rate tensor $\dot{\varepsilon}_{ij}$ is specified, according to which the centres of all the spheres in the periodic cell move, as though they are points in a continuum, to satisfy the equation

$$\Delta x_i = \dot{\varepsilon}_{ij} x_j \Delta t \tag{8.28}$$

in which x_j are the coordinates of a sphere centre and Δt is the small timestep used to advance the evolution of the system. Additional incremental displacements occur as a result of the interactions between contiguous spheres, as explained in Sect. 2.1. In order to permit quasi-static ($\dot{\varepsilon} = 10^{-5}$) simulations to be continued to large strains ($\varepsilon = 0.5$) within a reasonable timescale it is necessary to use density scaling. In the simulations reported the particle density is scaled up by a factor of 10^{12}. This increases the timestep from microseconds to seconds and does not affect the forces, displacements, work or energy.

In order to follow desired stress paths, servo-control algorithms are required and take the general forms

$$\dot{\varepsilon} = g(\sigma^* - \sigma) \tag{8.29a}$$

$$\dot{\varepsilon}^t = \dot{\varepsilon}^{t-\Delta t} + g(\sigma^* - \sigma) \tag{8.29b}$$

where σ^* is the desired value of stress, σ is the calculated value and g is a gain parameter whose appropriate value is obtained by trial and error. Equation (8.29a) is used to bring the system to equilibrium at a desired stress state, for example when performing isotropic compression. Equation (8.29b) is used to adjust the strain-rate to minimise the difference between the desired and calculated stress states when following a desired stress path, for example when performing shear deformation with the mean stress held constant. Both equations may be expressed in terms of individual components or combinations of components of the strain-rate and stress tensors depending on the desired loading path to be followed. More than one servo-control algorithm may be used but care must be taken to avoid conflicting adjustments to the strain-rate tensor.

To start a simulation, spheres are randomly generated within a cuboidal cell sufficiently large to provide an initial concentration of about 0.5 with no interparticle contacts. After generation, the system is subjected to isotropic compression using a strain-rate of 10^{-4}/s until the mean stress has reached a value of about 10 kPa. Isotropic compression is then continued using the servo-control algorithm defined by Eq. (8.29a), with a limit set to 10^{-4}/s, to raise the mean stress

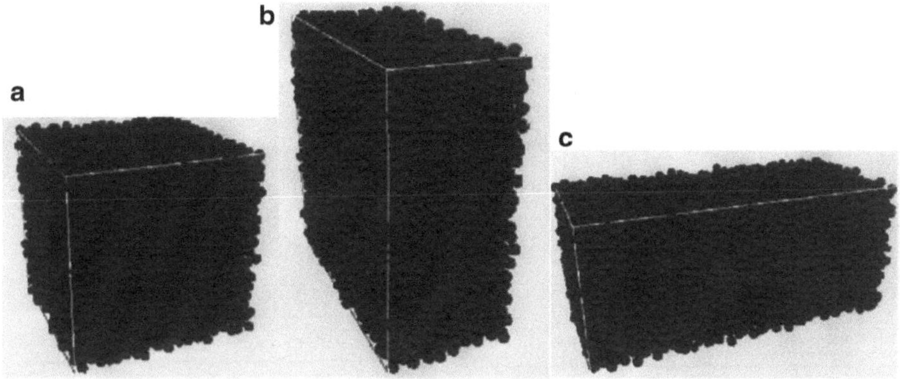

Fig. 8.29 Periodic cell under (**a**) isotropic compression (**b**) axisymmetric compression and (**c**) axisymmetric extension conditions

incrementally to the desired final value. If the value of interparticle friction to be used in the shear stage is introduced at the start of isotropic compression then a medium dense sample will be obtained. In order to obtain a dense sample the interparticle friction is set to zero until the mean stress reaches 90 % of the desired final value and then the desired value of interparticle friction is introduced for the final increment of mean stress. A very loose sample can be obtained by using the desired value of interparticle friction and switching off any particle rotation until the mean stress reaches 90 % of the desired final value. Using these procedures ensures that, with a sufficiently large number of particles, the prepared sample is isotropic in terms of both the stress tensor and the fabric tensor. Figure 8.29 illustrates a polydisperse system of 8000 spheres in a periodic cell under isotropic compression, axisymmetric compression and axisymmetric extension conditions.

8.3.1 Axisymmetric Compression

Figure 8.30 shows results obtained for a polydisperse system of 3620 elastic spheres during axisymmetric compression $(\sigma_1 > \sigma_2 = \sigma_3)$ with the mean stress, $p = (\sigma_1 + \sigma_2 + \sigma_3)/3$, maintained constant at 100 kPa. Results are shown for both a dense system and a loose one. Details of the particle properties etc. can be found in Thornton (2000).

Figure 8.30a shows the evolution of the deviator stress $(\sigma_1 - \sigma_3)$ with deviator strain $(\varepsilon_1 - \varepsilon_3)$ and the corresponding evolution of the volumetric strain $(\varepsilon_1 + \varepsilon_2 + \varepsilon_3)$ is shown in Fig. 8.30b. These two figures demonstrate that, qualitatively, the stress-strain-dilation response obtained for both the dense and loose systems is typical of that obtained in laboratory experiments. The initial shear modulus is much higher for the dense system, which exhibits a peak in the stress-

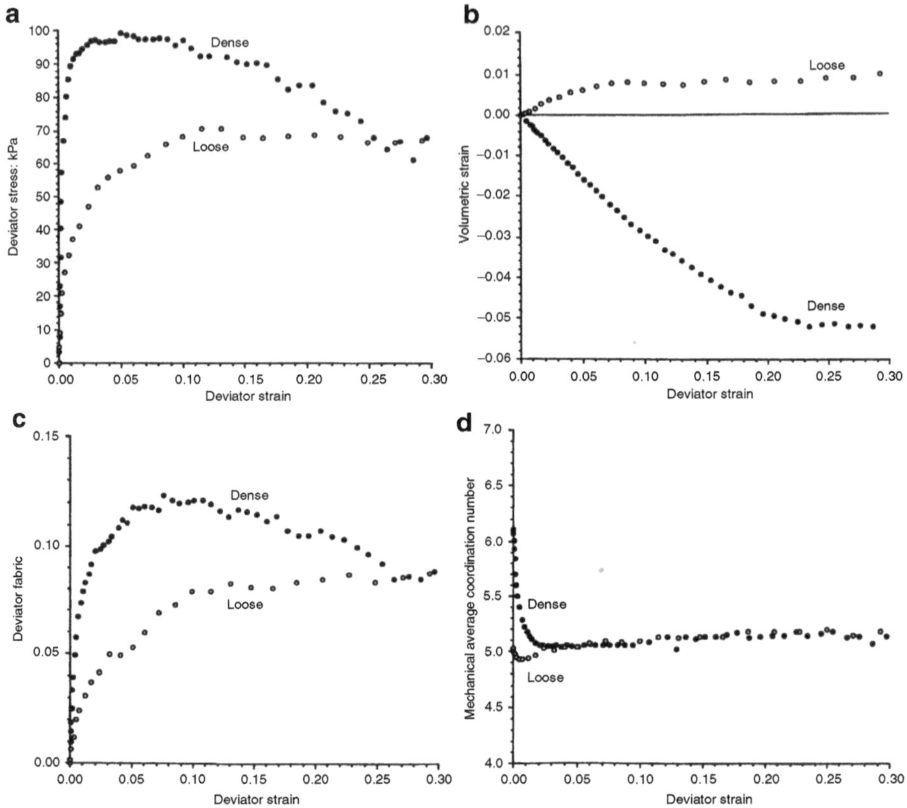

Fig. 8.30 Evolution of (**a**) deviator stress (**b**) volumetric strain (**c**) deviator fabric and (**d**) mechanical coordination number

strain curve at about 5 % strain followed by strain-softening behaviour. The loose system does not exhibit any strain-softening; the deviator stress increases at a decreasing rate until an essentially constant value is reached at about 15 % strain. The volumetric strain responses show that the dense system expands and the loose system contracts. At large strains, both systems deform at constant volume and this is associated with a constant deviator stress that is independent of the initial packing density. It is worth noting that the stress-strain response of the dense system demonstrates that strain-softening is a genuine material behaviour and not necessarily the result of non-uniform deformation due to the existence of platen boundaries, which can be the case in laboratory experiments.

It is now well established that shear deformation of compact particle systems produces an induced structural anisotropy that is developed primarily as a result of contact separation occurring in directions that are approximately orthogonal to the major principal strain direction. Structural anisotropy is defined by the distribution of contact orientations and characterised by a fabric tensor ϕ_{ij} defined by Eq. (2.25).

Consequently, in axisymmetric compression, the degree of structural anisotropy can conveniently be defined by the deviator fabric $(\phi_1 - \phi_3)$, which is plotted in Fig. 8.30c. The figure shows that the structural anisotropy increases at a decreasing rate to a maximum value that is dependent on the initial packing density. The dense system exhibits a decrease in structural anisotropy at strains in excess of 10 % until, at large strains, the deviator fabric is the same for both systems. Although there is a similarity between the stress-strain and fabric-strain curves shown in Fig. 8.30 there is no simple relationship between the two, as will be demonstrated later. The evolution of the mechanical coordination number Z_m, defined by Eq. (2.22), is shown in Fig. 8.30d. During the initial 3 % deviator strain, there is a rapid change in the mechanical coordination number until a 'critical' value is attained, which remains essentially constant thereafter, irrespective of whether the system is expanding or contracting. It is considered that this 'critical' value in some way reflects an underlying physical stability requirement but an exact explanation is still awaited.

8.3.1.1 Effect of Interparticle Friction

In experiments it is difficult to distinguish between the effects of contact friction and particle shape but, in numerical simulations the effects of contact friction can be isolated. Thornton and Sun (1993) reported simulations of axisymmetric compression using two different coefficients of interparticle friction $\mu = 0.3$ and $\mu = 0.6$ for both a dense system and a loose system. The results showed that an increase in μ resulted in an increase in shear modulus and shear strength for both systems and also resulted in a higher degree of structural anisotropy and higher rates of dilation. Further effects of interparticle friction are shown in Fig. 8.31.

Figure 8.31a illustrates the effect of interparticle friction on the mobilised shear strength $sin\phi = (\sigma_1 - \sigma_3)/(\sigma_1 + \sigma_3)$. Two sets of data obtained from the simulations are shown, corresponding to the peak value $sin\phi_{max}$ and the value at the end of the tests when the systems were deforming at constant volume $sin\phi_{cv}$, together with experimental measurements of $sin\phi_{cv}$ (including error bars) reported by Skinner (1969). There is reasonable agreement between the simulated and experimental data, except when $\mu \rightarrow 0$. Skinner's (1969) data suggest that $sin\phi_{cv}$ is independent of μ. This is contradicted by the results of the simulations, which are more convincing, since random assemblies of frictionless spheres are inherently unstable at all contacts all of the time, making it very difficult to develop any stable force transmission through the system that would lead to the development of a deviator stress. Also superimposed on the figure is the theoretical relationship suggested by Horne (1969). The significant difference between the theoretical prediction and the simulated data arises from the fact that the theory ignores the possibility of particle rotation. If particle rotation is prohibited then the simulation data may approach the theoretical curve.

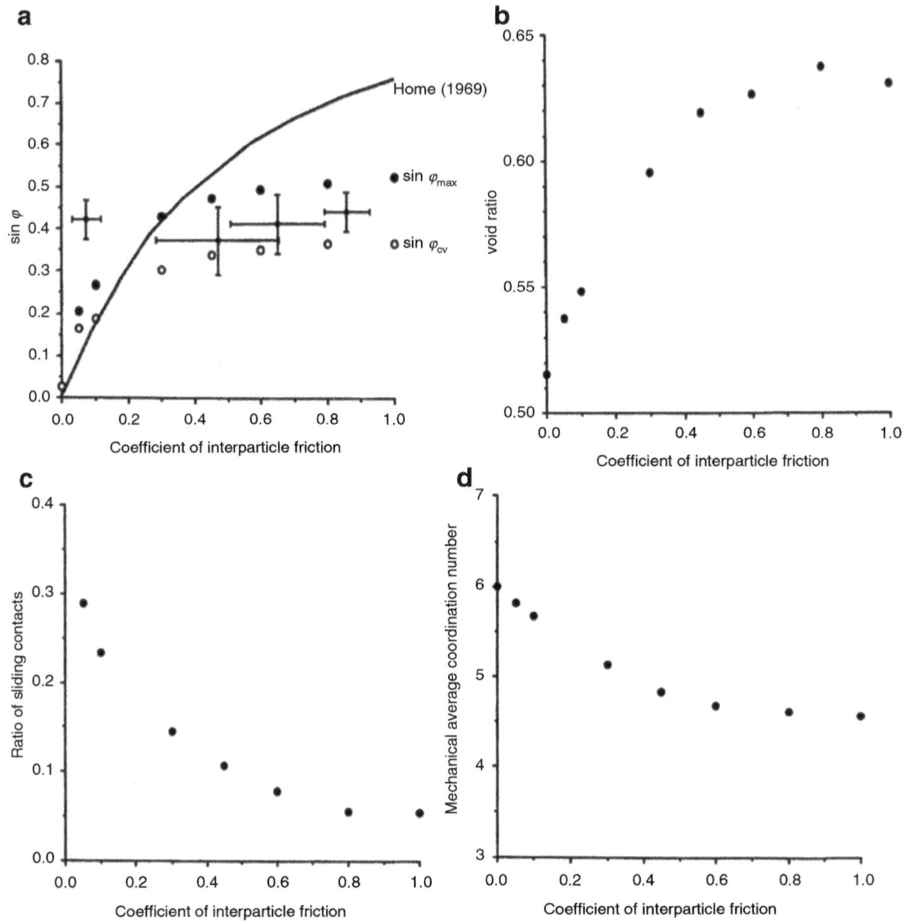

Fig. 8.31 Effect of interparticle friction on the (**a**) angle of shearing resistance (**b**) steady state void ratio (**c**) ratio of sliding contacts and (**d**) mechanical coordination number

In Fig. 8.31b the void ratio at large strain, when the systems were deforming at constant volume, is plotted against the interparticle friction coefficient. The figure shows that, for a given mean stress, the critical void ratio depends on the interparticle friction. In the case of $\mu = 0$ the system, like a liquid, did not change volume during shear. During all shear simulations, the ratio of sliding contacts increases rapidly to a value that thereafter remains constant throughout the shear stage. Figure 8.31c shows that increasing the interparticle friction reduces the ratio of sliding contacts, as would be expected. In Fig. 8.31d it can be seen that the critical mechanical coordination number decreases when the interparticle friction is increased.

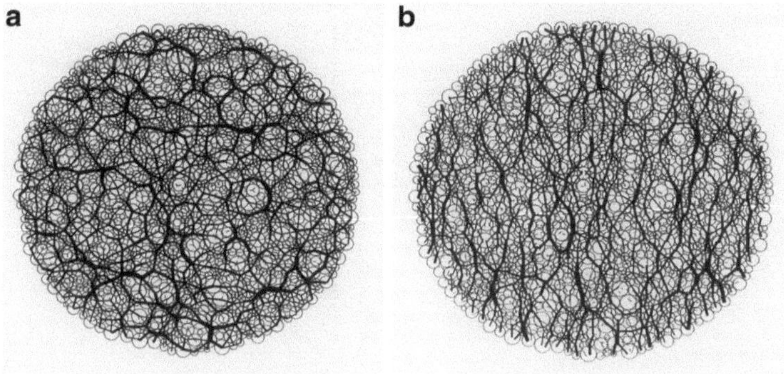

Fig. 8.32 Contact force transmission in a 2D array of polydisperse discs (**a**) isotropic stress state (**b**) deviator stress state

8.3.1.2 Significance of the Strong Force Chains

For any system of discrete particles subjected to external loading, the transmission of force from one boundary to another can only occur via the interparticle contacts. Intuitively, therefore, we expect that the distribution of contacts will determine the distribution of contact forces and that the forces will not necessarily be distributed uniformly. From both photoelastic studies of two-dimensional arrays of discs (Dantu 1957; Wakabayshi 1957; de Josselin de Jong and Verruijt 1969; Dresher and de Josselin de Jong 1972; Oda and Konishi 1974) and both 2D and 3D numerical simulations (Cundall and Strack 1979; Thornton and Barnes 1986; Radjai et al. 1997; Thornton 1997) it has been demonstrated that the applied load is largely transmitted by relatively rigid, heavily stressed chains of particles forming a relatively sparse percolating network of above-average contact forces. Groups of particles separating the strong force chains are only lightly loaded. The implication is that, in a random system of particles, the applied load will search for the shortest and most direct transmission path.

Even when both the microstructure and the stress state are isotropic, as can be seen in Fig. 8.32a, some contacts transmit forces several times those of others but with no preferred direction for the larger contact forces. During shear, the large forces immediately realign to become oriented in the direction of the major principal stress, as shown in Fig. 8.32b.

Rather than focus on the particles in the strong force chains, Radjai et al. (1997) suggested that the contact network may be partitioned into two complementary sub-networks: a 'strong' percolating sub-network of contacts transmitting above-average contact normal forces and a 'weak' sub-network of contacts transmitting below-average contact normal forces. From 2D simulations of biaxial compression on a polydisperse system of about 4000 rigid discs, they concluded that the forces in the strong sub-network account for all of the deviator stress, whereas the weak sub-network contributes only to the isotropic component of the stress tensor. They

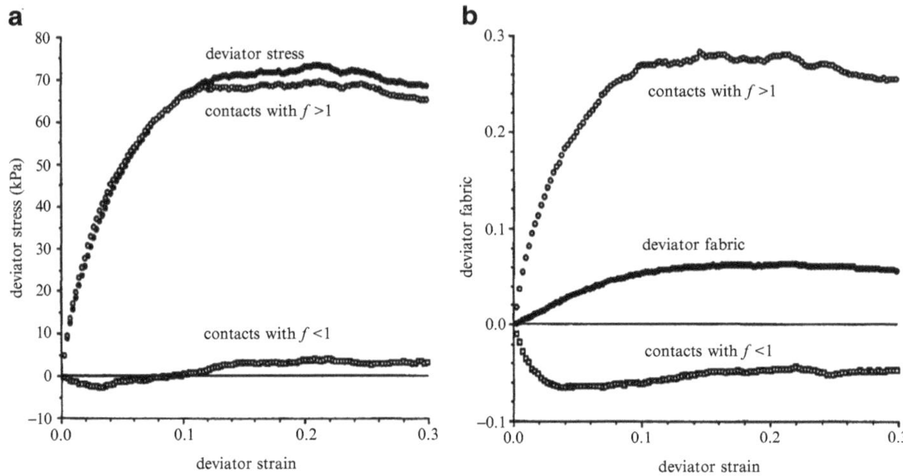

Fig. 8.33 Contributions of the weak ($f < 1$) and strong ($f > 1$) contact force sub-networks to the (**a**) deviator stress (**b**) deviator fabric

also found that the orientation of the induced structural anisotropy in the strong network coincided with the orientation of the stress tensor but the orientation of the induced structural anisotropy in the weak network was orthogonal to that of the strong network. This was also demonstrated for a 3D polydisperse system of 8000 elastic spheres by Thornton and Antony (1998).

The stress tensor and the fabric tensor are defined by Eqs. (2.45) and (2.25) respectively. Both of these tensors can be calculated on the basis of the separate contributions of the contacts transmitting larger than average contact forces and the contacts transmitting less than average contact forces, i.e.

$$\sigma_{ij} = \sigma^s_{ij} + \sigma^w_{ij} \text{ and } \phi_{ij} = q\phi^s_{ij} + (1 - q)\phi^w_{ij} \tag{8.30}$$

where the superscripts s and w indicate the strong and weak sub-networks and q is the proportion of contacts in the strong sub-network.

Figure 8.33a shows the evolution of $(\sigma_1 - \sigma_3)$, $(\sigma_1 - \sigma_3)^s$ and $(\sigma_1 - \sigma_3)^w$; the evolution of $(\phi_1 - \phi_3)$, $(\phi_1 - \phi_3)^s$ and $(\phi_1 - \phi_3)^w$ is shown in Fig. 8.33b. In both figures the symbol $f = N/\langle N\rangle$ where N is the normal contact force. It is clear that the results fully support the conjectures of Radjai et al. (1997). Throughout the simulation the percentage of contacts in the strong sub-network was about 33 %. The significance of this is that a system of compact particles is a highly redundant system but in order to mobilise shear resistance only a minority of the contacts are required as a consequence of the way in which forces are transmitted through the system. The consequence of this is that the system is highly adaptive and therefore robust. If the direction of principal stress is suddenly rotated then the force transmission immediately adapts and selects a new set of contacts to transmit the

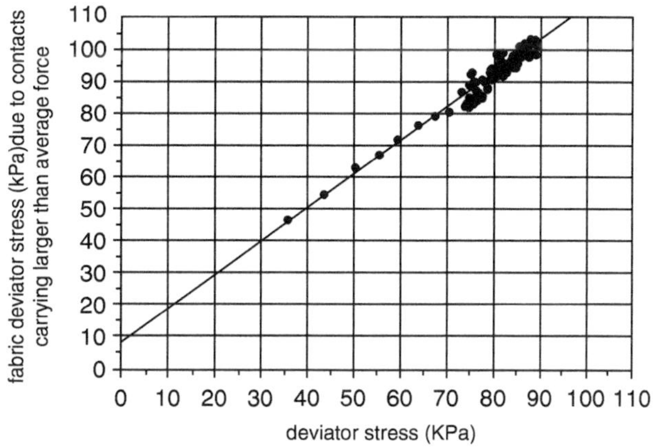

Fig. 8.34 Correlation between the deviator stress and the fabric deviator stress due to contacts transmitting larger than average forces

strong contact forces. Because of the high redundancy this is achieved without any major change in the microstructure.

Although the evolutions of the deviator stress and the deviator fabric are qualitatively similar, as shown in Fig. 8.30, there is no simple correlation between them. However, it has been shown above that, to a close approximation, the deviator stress is entirely due to the sub-network of favourably oriented contacts transmitting the larger than average contact forces. In order to examine this further, it is useful to define a fabric stress tensor

$$\sigma_{ij}^{f} = \sigma_{kk}\phi_{ij} \tag{8.31}$$

and to decompose the fabric tensor into the separate contributions of the strong and weak contacts. We may then examine the correlation between the fabric deviator stress due to the strong sub-network, $\sigma_{kk}(\phi_1 - \phi_3)^s$, and the mobilised deviator stress $(\sigma_1 - \sigma_3)$. This is shown in Fig. 8.34 and it can be seen that the correlation is excellent if the small initial value of fabric deviator stress is ignored.

8.3.1.3 Effect of Plastic Deformation at the Interparticle Contacts

Using the elastic-plastic contact force model described in Sect. 3.2, axisymmetric compression tests were simulated, with the mean stress held constant at 100 kPa, using a polydisperse system of 4000 spheres in order to examine the effect of plastic yield at the contacts. Three tests were simulated. In the first test a very high limiting contact pressure p_y was specified so that no contacts yielded in order to provide data for a system of elastic spheres for comparison. Using the results of the first test, the distribution of mean normal contact pressures was examined prior to shear. From this information a limiting contact pressure was selected such that about 10 % of

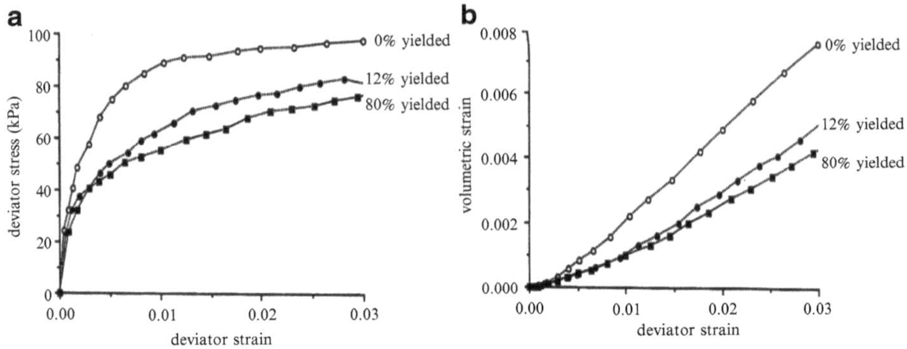

Fig. 8.35 Effect of plastic deformation at the contacts on the (**a**) deviator stress (**b**) volumetric strain

contacts would have yielded if the spheres had not been elastic. Starting from a lower isotropic stress state of 50 kPa the calculated limiting contact pressure was specified and the system was then isotropically compressed to 100 kPa to provide a second specimen to be sheared at constant mean stress. When checked, it was found that in fact 12 % of contacts had deformed plastically before the start of the shear stage. A third specimen was also prepared in the same manner but with a lower limiting contact pressure specified, which resulted in plastic deformation at 80 % of the contacts prior to shear.

It can be seen from Fig. 8.35 that plastic deformation at interparticle contacts has a pronounced effect on the stress-strain-dilation behaviour. There is a large reduction in the shear modulus even when only a relatively small number of contacts have yielded. Increasing the number of yielded contacts reduces the shear modulus more but not to a large degree. Volumetric expansion is greatest for the elastic sphere system. A small number of yielded contacts significantly reduces the rate of dilation, which is further reduced slightly when there are many contacts that have yielded. The explanation for the behaviour shown in Fig. 8.35 is that the magnitude of the deviator stress is dictated by the contact forces in the strong sub-network of contacts transmitting greater than average forces, as shown in Fig. 8.33a. Generally there are 30–35 % of contacts in the strong force sub-network. Consequently, the contacts that had yielded in the second test (12 %) were all in the strong sub-network. It can also be inferred that the results obtained for the specimen with 80 % yielded contacts is very similar to what would have been obtained if all the contacts in the strong sub-network had yielded and there were no yielded contacts in the weak sub-network.

8.3.1.4 Elastic Properties

The elastic response of granular media has been examined by Digby (1981), Walton (1987) and Bathurst and Rothenburg (1988). Subsequent work using two

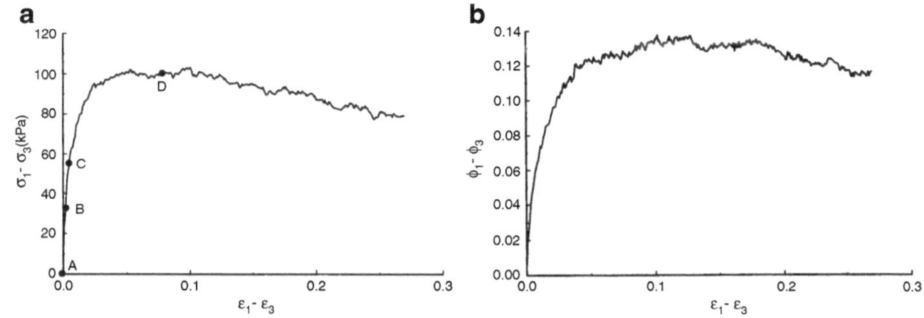

Fig. 8.36 Evolution of (**a**) deviator stress (**b**) deviator fabric

homogenization techniques has suggested that the kinematic hypothesis, which assumes a uniform strain field, provides an upper bound to the elastic moduli and the static hypothesis, which assumes a uniform stress field, provides a lower bound, see Kruyt and Rothenburg (2002) or Hicher and Chang (2005) for details. However, both pairs of researchers compare their predictions with results of DEM simulations of irrotational bonded spheres which they consider provides the true solution. In the opinion of the author this is not exactly correct.

If one applies a small deformation to a system of particles then, no matter how small the deformation, there will be a small change in the coordinates of all of the particles. Consequently, all contact normal vectors will experience a small change in their orientations resulting in a small, irreversible change in the fabric tensor. Therefore the elastic properties of particle systems can only be approximated by performing DEM simulations. In order to identify the true elastic properties it is essential that absolutely no deformation is actually applied.

As demonstrated in Sect. 2.2.4, the fourth order tensor S_{ijkl} defining the assembly modulus was derived and given by Eq. (2.60). The tensor is a function of the current distribution of contact normal vectors, the current normal and tangential contact stiffnesses and the current contact density. In DEM simulations all this information is known at any stage of a test. However, if a small incremental strain is applied, no matter how small, then this information will change due to irreversible, inelastic, deformation. Consequently, what will be obtained is the 'small strain stiffness' not the true elastic properties.

Figure 8.36 shows the evolution of the deviator stress and deviator fabric during axisymmetric compression of a polydisperse system of 3600 elastic spheres performed in a periodic cell with the mean stress held constant at 100 kPa. The properties of the particles were: Young's modulus = 70 GPa, Poisson's ratio = 0.3, interparticle friction coefficient = 0.5 and sizes ranging from 30 μm to 90 μm.

For every point on the pre-peak loading curve shown in Fig. 8.36a all the components of the fourth order tensor S_{ijkl} were calculated. However, it is then

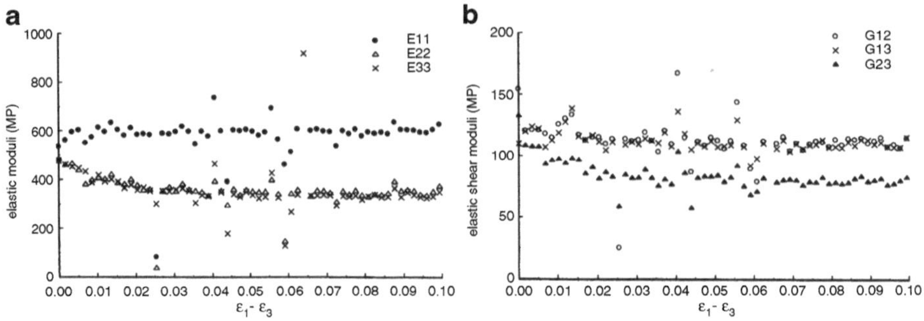

Fig. 8.37 Evolution of elastic moduli (**a**) E_{11}, E_{22} and E_{33} (**b**) G_{12}, G_{13} and G_{23}

convenient to calculate the compliance tensor $C_{ijkl} = S_{ijkl}^{-1}$ in order to compare directly with the classical anisotropic elastic compliance matrix given by

$$\begin{bmatrix} \frac{1}{E_{11}} & \frac{-\nu_{21}}{E_{22}} & \frac{-\nu_{31}}{E_{33}} & 0 & 0 & 0 \\ \frac{-\nu_{12}}{E_{11}} & \frac{1}{E_{22}} & \frac{-\nu_{32}}{E_{33}} & 0 & 0 & 0 \\ \frac{-\nu_{13}}{E_{11}} & \frac{-\nu_{23}}{E_{22}} & \frac{1}{E_{33}} & 0 & 0 & 0 \\ 0 & 0 & 0 & \frac{1}{2G_{23}} & 0 & 0 \\ 0 & 0 & 0 & 0 & \frac{1}{2G_{13}} & 0 \\ 0 & 0 & 0 & 0 & 0 & \frac{1}{2G_{12}} \end{bmatrix} \tag{8.32}$$

The evolution of the elastic parameters, obtained as described above, are shown in Figs. 8.37 and 8.38.

From Fig. 8.37a it can be seen that the modulus E_{11} in the direction of the major principal stress/strain is greater than the moduli in the other two directions. This demonstrates the anisotropic elasticity of the system. The agreement between E_{22} and E_{33} demonstrates that the elasticity is transversely isotropic. The moduli E_{22} and E_{33} decrease at a decreasing rate until a constant value is reached before peak shear strength is reached. The reduction in the values is due to the loss of contacts in the 2- and 3-directions. The value of E_{11} remains constant after the initial shearing stage. The initial increase in E_{11} is not due to an increase in the number of contacts in the one-direction but due to an increase in contact stiffness (Hertzian) when the large force chains quickly realign with the σ_1 direction. Ignoring the random fluctuations, it is clear from Fig. 8.37b that, after the initial 2 % strain, the three shear moduli remain constant and that $G_{12} = G_{13} > G_{23}$.

Figure 8.38 shows that $\nu_{ij} \neq \nu_{ji}$ except for $\nu_{23} = \nu_{32}$. This is necessary to ensure symmetry of C_{ijkl}. It can be seen from Fig. 8.38 that the average values approximate to $\nu_{12} = \nu_{13} = 0.10$, $\nu_{21} = \nu_{31} = 0.05$ and $\nu_{23} = \nu_{32} = 0.075$. These values are in

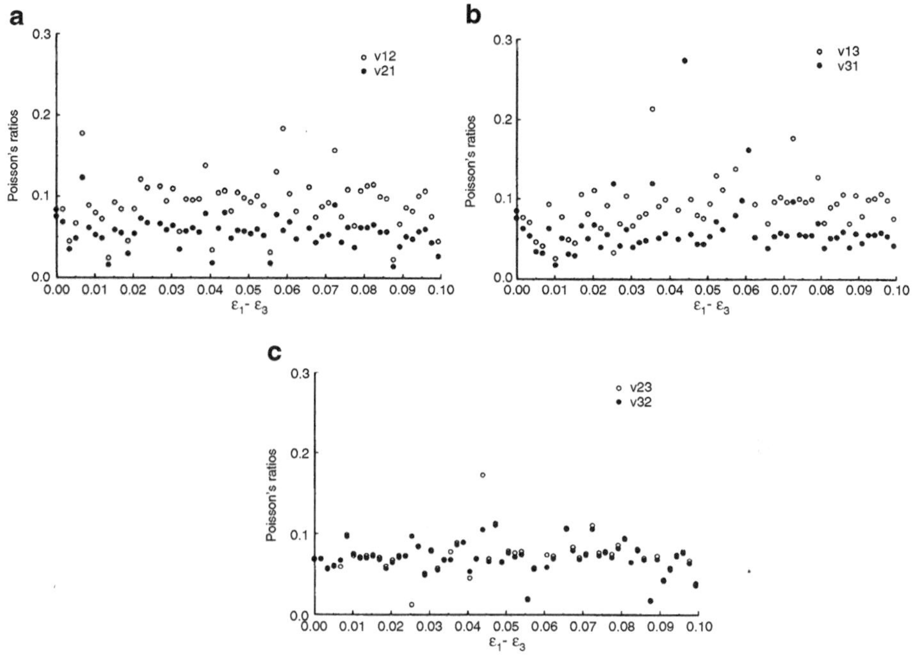

Fig. 8.38 Evolution of Poisson's ratios (**a**) ν_{12} and ν_{21} (**b**) ν_{13} and ν_{31} (**c**) ν_{23} and ν_{32}

reasonable agreement with values predicted by Bathurst and Rothenburg (1988) for random arrays of equal-sized spheres with assumed linear contact stiffnesses. Further details can be found in Thornton and Zhang (2005).

8.3.2 General 3D Stress States

Most 3D simulations of element tests are axisymmetric compression. This is a state of stress that is rarely encountered in real world situations. The general case is one in which $\sigma_1 \neq \sigma_2 \neq \sigma_3$. In this general case Mohr stress circles are not useful and, instead, one has to consider principal stress space in which the state of stress is given by a point and represented by a vector σ from the origin, as illustrated in Fig. 8.39a. The stress vector has two orthogonal components: an isotropic stress vector

$$\sigma_z = \frac{\sigma_1 + \sigma_2 + \sigma_3}{\sqrt{3}} \quad (8.33)$$

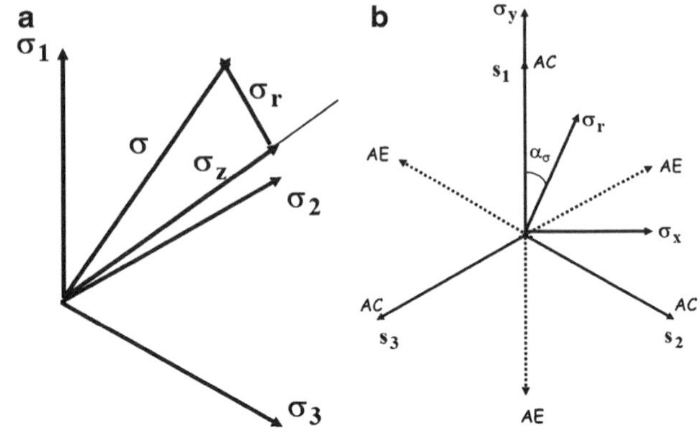

Fig. 8.39 Principal stress space (**a**) general view (**b**) deviatoric plane

and a deviatoric stress vector

$$\sigma_r = \sqrt{\sigma_y^2 + \sigma_x^2} \tag{8.34}$$

where the σ_y and σ_x axes are illustrated in Fig. 8.39b, which shows the deviatoric plane as viewed along the z (isotropic) axis, and defined by

$$\sigma_y = \frac{2\sigma_1 - \sigma_2 - \sigma_3}{\sqrt{6}} \text{ and } \sigma_x = \frac{\sigma_2 - \sigma_3}{\sqrt{2}} \tag{8.35}$$

It should be noted that, in principal stress space, the subscripts 1, 2 and 3 indicate the principal stresses in the three orthogonal directions of the global reference frame, as shown in Fig. 8.39a, and not necessarily the major, intermediate and minor principal stresses. This convention also applies to the strain and fabric tensors considered later. Figure 8.39b indicates the three possible directions for axisymmetric compression (AC) and axisymmetric extension (AE). For general states of stress, the orientation of the deviatoric stress vector σ_r is defined by the Lode angle α_σ, as indicated in Fig. 8.39b. Taking $\alpha_\sigma = 0$ in the 1-direction, the Lode angle is defined by

$$tan\,\alpha_\sigma = \frac{\sqrt{3}(\sigma_2 - \sigma_3)}{2\sigma_1 - \sigma_2 - \sigma_3} \tag{8.36}$$

In traditional soil mechanics, shear strength is defined by the Mohr-Coulomb criterion, $\sin\phi$ = constant, where ϕ is the inclination of the line tangent to a set of Mohr stress circles. In principal stress space, all points representing failure states of stress lie on a failure surface (or limit surface) that is conical, as shown in Fig. 8.40a for axisymmetric compression and axisymmetric extension. Figure 8.40b shows various possible failure criteria, as viewed on the deviatoric stress plane, including Mohr-Coulomb ($\sin\phi$ = constant) extended Tresca ($\sigma_1 - \sigma_3$ = constant), and extended von Mises (σ_r = constant) as considered by Bishop (1966) plus more recently proposed forms suggested by Matsuoka and Nakai (1974) and Lade and Duncan (1975).

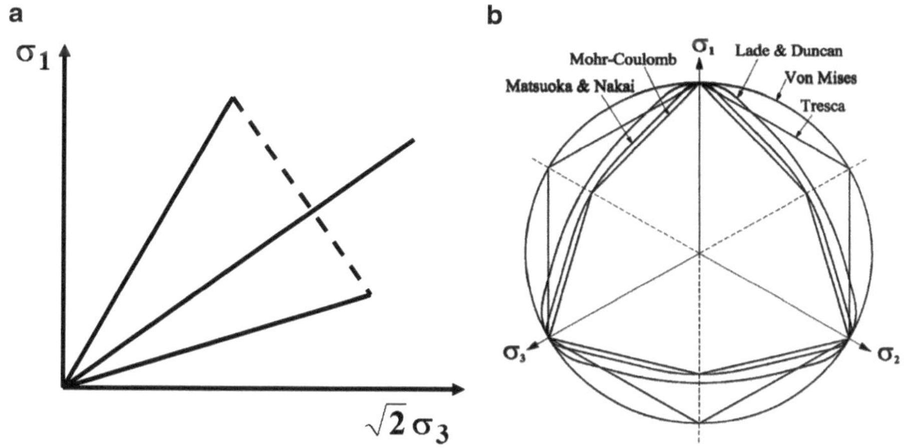

Fig. 8.40 Failure surfaces on the (**a**) axisymmetric plane (**b**) deviatoric plane

Fig. 8.41 Deviatoric failure surfaces for a dense and a loose specimen

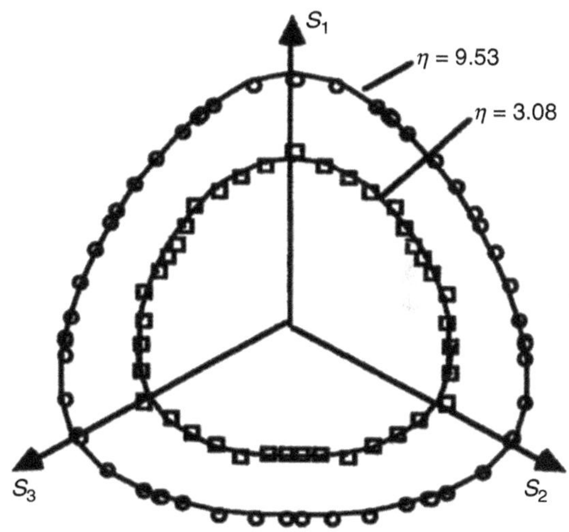

Figure 8.41 shows the deviatoric failure states of stress reported by Thornton (2000) for both a dense and a loose specimen. Both systems were sheared with the mean stress held constant and subjected to deviatoric loading during which the ratio

$$b = \frac{\sigma_2 - \sigma_3}{\sigma_1 - \sigma_3} \tag{8.37}$$

was maintained constant using a servo-control algorithm like Eq. (8.29b) with σ replaced by b. A series of simulations covering the range $0 < b < 1$ were performed ranging from axisymmetric compression ($b = 0$) to axisymmetric extension ($b = 1$). For each value of b, the state of stress corresponding to failure (maximum deviatoric stress) was identified and plotted on the deviatoric stress plane. In all cases the principal stress in the one-direction was the major principal stress. Then, assuming a sixfold symmetry, the complete failure surfaces were produced as shown in the figure. Superimposed on the figure is the failure criterion proposed by Lade and Duncan (1975), which is defined by

$$\eta = \frac{I_1^3}{I_3} - 27 \tag{8.38}$$

where $I_1 = \sigma_1 + \sigma_2 + \sigma_3$ is the first stress invariant and $I_3 = \sigma_1\sigma_2\sigma_3$ is the third stress invariant. The figure shows excellent agreement between the simulation data and Eq. (8.38) which has also been confirmed by Calvetti et al. (2003); and by Ng (2004) for assemblies of ellipsoidal particles.

Rather than control a stress path it is easier to apply a strain path and monitor the stress response. Thornton and Zhang (2010) reported results obtained from radial deviatoric straining of a dense polydisperse system of 27,000 elastic spheres that had initially been isotropically compressed to 100 kPa. The deviatoric strain increment vector is defined by

$$\Delta\varepsilon_d = \frac{1}{\sqrt{3}}\left[(\Delta\varepsilon_1 - \Delta\varepsilon_2)^2 + (\Delta\varepsilon_1 - \Delta\varepsilon_3)^2 + (\Delta\varepsilon_2 - \Delta\varepsilon_3)^2\right]^{1/2} \tag{8.39}$$

The Lode angle for strain is defined as

$$tan\,\alpha_\varepsilon = \frac{\sqrt{3}(\Delta\varepsilon_2 - \Delta\varepsilon_3)}{2\Delta\varepsilon_1 - \Delta\varepsilon_2 - \Delta\varepsilon_3} \tag{8.40}$$

It follows from Eqs. (8.39) and (8.40) that

$$\Delta\varepsilon_2 - \Delta\varepsilon_3 = \sqrt{2}\Delta\varepsilon_d sin\,\alpha_\varepsilon \tag{8.41a}$$

$$2\Delta\varepsilon_1 - \Delta\varepsilon_2 - \Delta\varepsilon_3 = \sqrt{6}\Delta\varepsilon_d cos\,\alpha_\varepsilon \tag{8.41b}$$

$$\Delta\varepsilon_1 + \Delta\varepsilon_2 + \Delta\varepsilon_3 = 0 \tag{8.41c}$$

and therefore

$$\Delta\varepsilon_1 = \sqrt{\frac{2}{3}}\Delta\varepsilon_d \cos\alpha_\varepsilon \tag{8.42a}$$

$$\Delta\varepsilon_2 = \Delta\varepsilon_d\left(\frac{sin\,\alpha_\varepsilon}{\sqrt{2}} - \frac{cos\,\alpha_\varepsilon}{\sqrt{6}}\right) \tag{8.42b}$$

$$\Delta\varepsilon_3 = -\Delta\varepsilon_d\left(\frac{sin\,\alpha_\varepsilon}{\sqrt{2}} + \frac{cos\,\alpha_\varepsilon}{\sqrt{6}}\right) \tag{8.42c}$$

Consequently, if the simulations are to be performed at constant volume, for any prescribed values of $\Delta\varepsilon_d$ and α_ε the corresponding strain-rate tensor is

$$\dot{\varepsilon}_{ij} = \frac{1}{N\Delta t}\begin{bmatrix} \Delta\varepsilon_1 & 0 & 0 \\ 0 & \Delta\varepsilon_2 & 0 \\ 0 & 0 & \Delta\varepsilon_3 \end{bmatrix} \tag{8.43}$$

where N is the number of timesteps Δt chosen to reach the desired value of $\Delta\varepsilon_d$.

If the simulations are to be performed at constant mean stress then it is, in addition, necessary to use a servo-control of the form given in Eq. (8.29b) by which the volumetric strain-rate is continuously adjusted to maintain the trace of the stress tensor constant at the desired value, i.e.

$$\dot{\varepsilon}^{t}_{kk} = \dot{\varepsilon}^{t-\Delta t}_{kk} + g\left(\sigma^{*}_{kk} - \sigma_{kk}\right) \tag{8.44}$$

The adjusted volumetric strain-rate then needs to be distributed to the three normal strain-rates in a manner that prevents the servo-control from modifying the dilatancy ratios. This is achieved by the following equations

$$\dot{\varepsilon}^{t}_{11} = \dot{\varepsilon}^{t-\Delta t}_{11} + \dot{\varepsilon}^{t}_{kk}\left|\frac{\dot{\varepsilon}_{11}}{\dot{\varepsilon}_{kk}}\right|^{t-\Delta t} \tag{8.45a}$$

$$\dot{\varepsilon}^{t}_{22} = \dot{\varepsilon}^{t-\Delta t}_{22} + \dot{\varepsilon}^{t}_{kk}\left|\frac{\dot{\varepsilon}_{22}}{\dot{\varepsilon}_{kk}}\right|^{t-\Delta t} \tag{8.45b}$$

$$\dot{\varepsilon}^{t}_{33} = \dot{\varepsilon}^{t-\Delta t}_{33} + \dot{\varepsilon}^{t}_{kk}\left|\frac{\dot{\varepsilon}_{33}}{\dot{\varepsilon}_{kk}}\right|^{t-\Delta t} \tag{8.45c}$$

In a constant mean stress test, at each timestep, Eq. (8.44) and Eqs. (8.45a, b and c) are applied before Eq. (8.43). Since Eq. (8.43) involves no change in volumetric strain, see Eq. (8.41c), there are no conflicting adjustments made to the strain-rate tensor.

With the mean stress held constant, strain probes were applied for 10° increments of the Lode angle from 0 to 360°, as shown in Fig. 8.42.

At the end of each probe, the stress tensor was calculated and the corresponding stress state was plotted on the deviatoric plane of principal stress space. For a given value of deviatoric strain, defined by the circles in Fig. 8.42b, there is a corresponding surface in stress space, which is referred to as a stress response envelope. The stress response envelopes obtained for different values of deviatoric strain are shown in Fig. 8.43a in which the stresses are dimensionless by dividing by the mean stress. Superimposed on the figure is a set of Lade surfaces, defined by Eq. (8.38), fitted to the data points corresponding to the plane strain states. The agreement with all data sets is excellent. Figure 8.43a shows that Eq. (8.38) is not simply a failure criterion but defines the evolution of the mobilised strength, defined by the parameter η, throughout deformation. In the context of elastoplasticity theory, the stress response envelopes can be considered to be 'bounding surfaces', see Dafalias and Popov (1976).

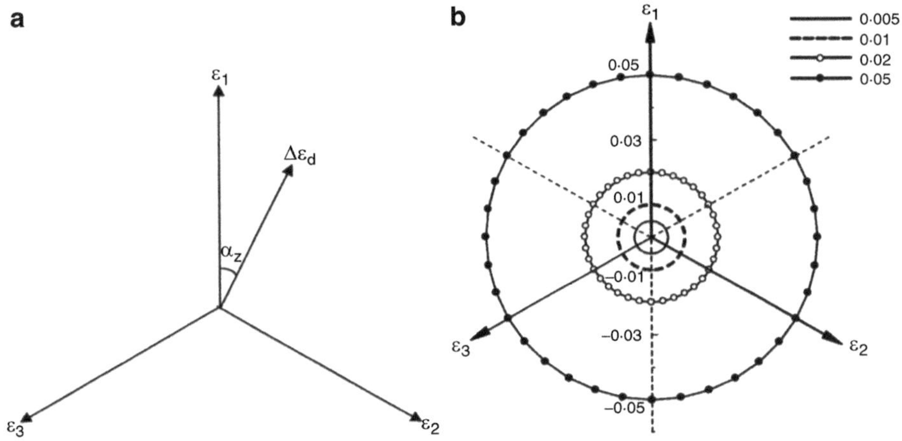

Fig. 8.42 Radial deviatoric strain probes (**a**) strain-increment vector (**b**) complete sets of probes for different values of deviatoric strain

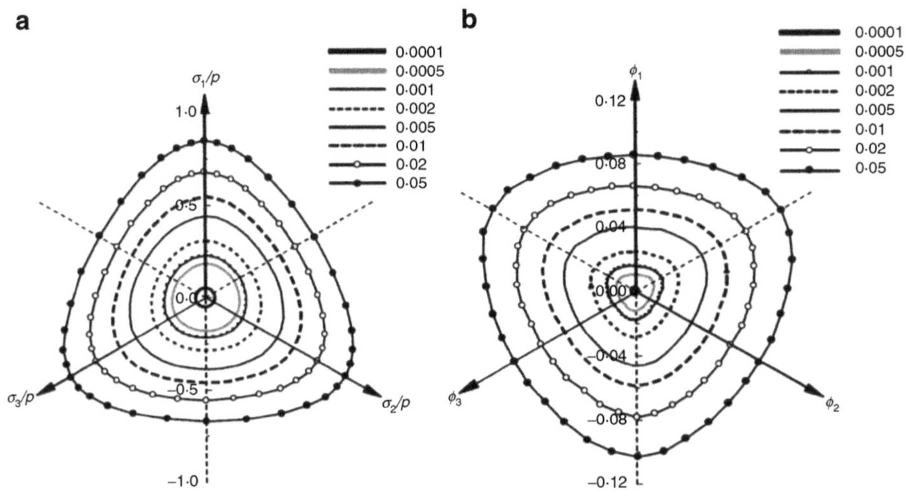

Fig. 8.43 Response envelopes for (**a**) stress (**b**) fabric

Since it is a tensor, the fabric tensor can be manipulated in the same way as the stress tensor. Consequently, for the deviatoric strain probes shown in Fig. 8.42b the corresponding fabric response envelopes can be plotted in principal fabric space, as shown in Fig. 8.43b. Superimposed on Fig. 8.43b is a set of surfaces that take the shape of 'inverted' Lade surfaces that can be characterised by the parameter η^* where

$$\eta^* = \frac{I_1^3}{2I_1I_2 - 3I_3} \tag{8.46}$$

In terms of the fabric tensor

$$I_1 = \phi_1 + \phi_2 + \phi_3 = 1 \tag{8.47a}$$

$$I_2 = \phi_1\phi_2 + \phi_1\phi_3 + \phi_2\phi_3 \tag{8.47b}$$

$$I_3 = \phi_1\phi_2\phi_3 \tag{8.47c}$$

Consequently, the fabric response envelopes can be characterised by the parameter η_f^* defined by

$$\eta_f^* = \frac{1}{2I_2 - 3I_3} \tag{8.48}$$

and it can be seen in Fig. 8.43b that the curves defined by Eq. (8.48) show excellent agreement with the data sets.

In soil mechanics it is common to use the parameter $\sin\phi = (\sigma_1 - \sigma_3)/(\sigma_1 + \sigma_3)$ to define the mobilised shear strength. Figure 8.44a shows the evolution of $\sin\phi$ with deviatoric strain for a separate series of simulations on the same system in which radial deviatoric straining was performed for different constant values of the parameter $b_\varepsilon = (\varepsilon_2 - \varepsilon_3)/(\varepsilon_1 - \varepsilon_3)$. Note that a constant b_ε test is equivalent to a constant Lode angle for strain since $tan\,\alpha_\varepsilon = \sqrt{3}b_\varepsilon/(2 - b_\varepsilon)$.

It is clear from Fig. 8.44a that $\sin\phi$ is very sensitive to strain path. In terms of shear strength, as defined by $\sin\phi_{\max}$, the lowest value occurs in axisymmetric compression ($b_\varepsilon = 0$) and the highest value occurs in plane strain ($b_\varepsilon = 0.5$). The second lowest value shown in the figure corresponds to axisymmetric extension ($b_\varepsilon = 1$). Figure 8.44b shows the evolution of volumetric strain with deviatoric strain, for the complete range of b_ε values. It can be seen that, at least until well past the peak strength, all the data sets collapse on to one master curve.

Figure 8.44c shows that if the evolution of the Lade parameter η is plotted against deviatoric strain then all the data sets collapse on to one master curve. This demonstrates that, rather than $\sin\phi_{\max}$, the general definition of shear strength is given by $\eta_{\max}$. Since the fabric response envelopes are inverted Lade surfaces, Fig. 8.43b, it can be anticipated that if the characteristic fabric parameter, defined by Eq. (8.48), is plotted against deviatoric strain then all data sets will also collapse on to one master curve. This is confirmed in Fig. 8.44d allowing for the exaggerated differences due to the vertical scale selected.

In Sect. 8.3.1.2 the significance of the strong force chains was discussed and it was demonstrated that the deviator stress is almost entirely due to the contact forces in the strong sub-network of contacts transmitting greater-than-average forces, as shown in Fig. 8.33a. The fabric response envelope for $\varepsilon_d = 0.05$, shown in Fig. 8.43b, is replotted in Fig. 8.45a. Superimposed on the figure is the fabric response envelope for the strong sub-network which can be seen to take the form of a Lade surface. By multiplying the fabric tensor by the trace of the stress tensor a fabric stress tensor is obtained, as given by Eq. (8.31). Dividing the fabric stress tensor by the mean stress

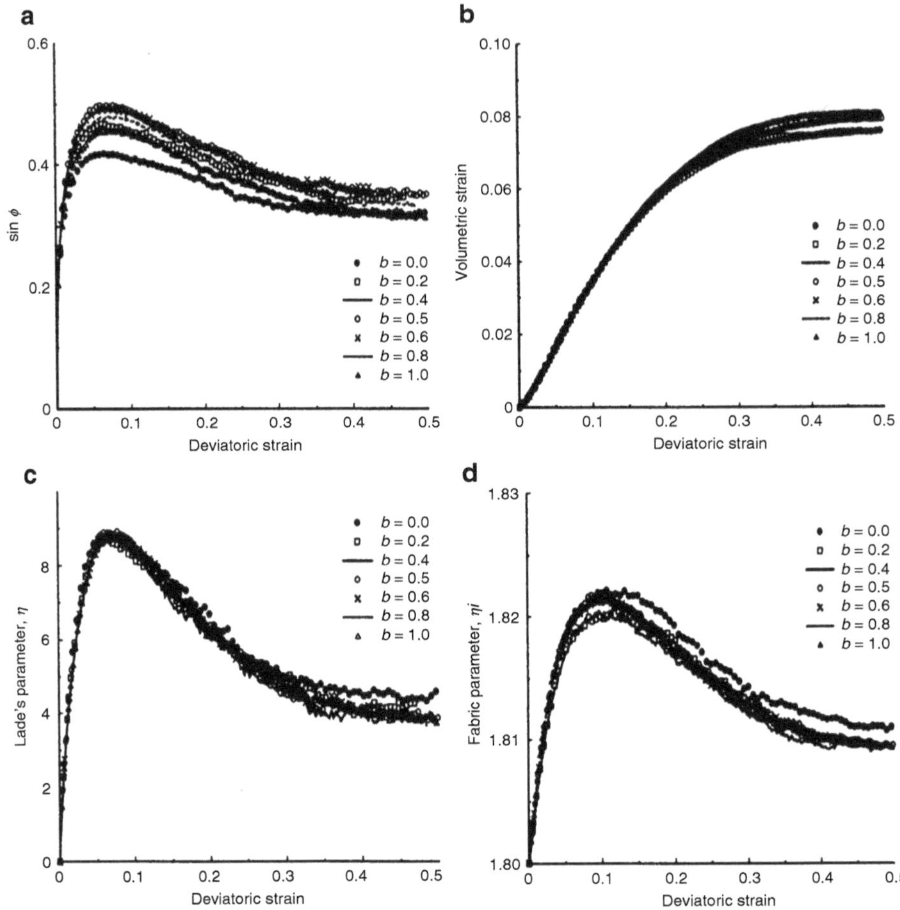

Fig. 8.44 Evolution of (**a**) sinϕ (**b**) volumetric strain (**c**) η and (**d**) η_f^*

the fabric response envelope for the strong sub-network can be obtained, as shown in Fig. 8.45b. Superimposed on Fig. 8.45b is the corresponding total stress response envelope. The figure shows that there is excellent agreement between the stress response envelope and fabric stress response envelope for the strong sub-network, except in the regions close to the axisymmetric compression states. Figure 8.45b is, in effect, a generalised extension to Fig. 8.34.

A problem with the Lade parameter η is that it can in theory range from zero to infinity and it is not clear whether any value, say 15, indicates a high or very high strength. Consequently, an alternative parameter is sought. Rearranging Eq. (8.38)

$$\frac{I_3}{I_1^3} = \frac{1}{27 + \eta} \tag{8.49}$$

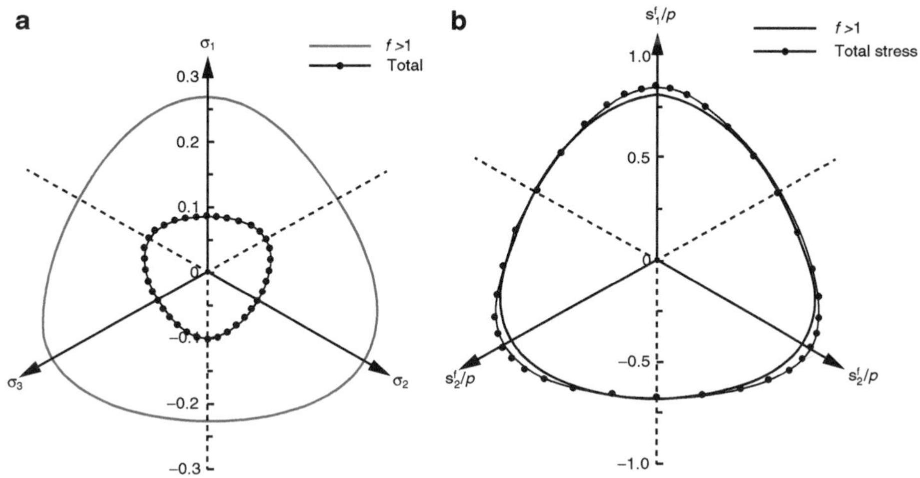

Fig. 8.45 Response envelopes (**a**) fabric (**b**) stress and fabric stress

or

$$I_3^* = \frac{\sigma_1 \sigma_2 \sigma_3}{p\ p\ p} = \frac{27}{27+\eta} \tag{8.50}$$

where I_3^* is the third invariant of the normalised stress tensor, in which the stresses are normalised by dividing by the mean stress. Unfortunately, the value of this parameter reduces with increase in the size of the stress response envelope. Consequently, in order to define the size of a stress response envelope the following parameter is used

$$\left(1 - I_3^*\right) = \frac{\eta}{27+\eta} \quad 0 \leq \left(1 - I_3^*\right) \leq 1 \tag{8.51}$$

The lower limit of zero corresponds to an infinitesimally small circle and the upper limit of unity corresponds to an equilateral triangle that defines the limit of compressive stress space on the deviatoric plane.

A close examination of Figs. 8.42b and 8.43a reveals that for any radial strain path, except axisymmetric strain states, the corresponding stress paths are curved. It is therefore of interest to relate the Lode angle for stress, defined by Eq. (8.36), to the Lode angle for strain, defined by Eq. (8.40). This is illustrated in Fig. 8.46. For axisymmetric compression $\alpha_\varepsilon = 0°$, 120° or 240° and $\cos(3\alpha_\sigma) = \cos(3\alpha_\varepsilon) = 1$ and for axisymmetric extension $\alpha_\varepsilon = 60°$, 180° or 300° and $\cos(3\alpha_\sigma) = \cos(3\alpha_\varepsilon) = -1$.

For non-axisymmetric strain states the data points in Fig. 8.46a are fitted by circular arcs that are characterised by the curvature $1/R$ which, since $\cos a \leq 1$, has

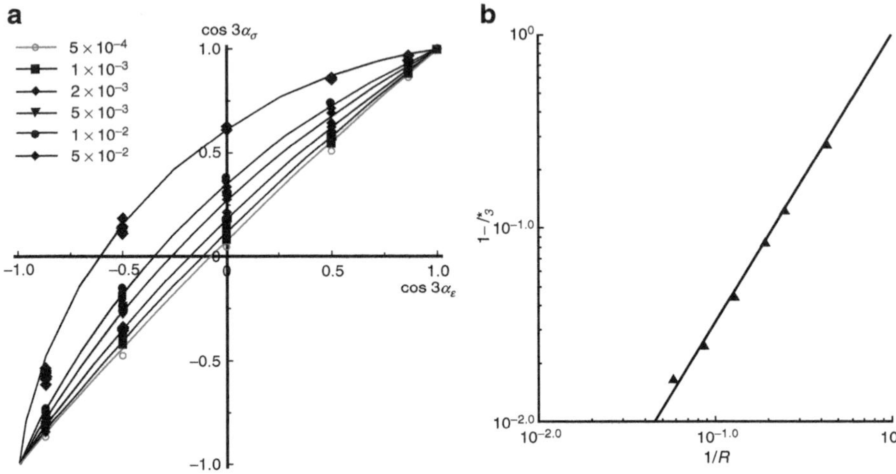

Fig. 8.46 Flow rule (**a**) and scaling law (**b**)

limits $0 \leq 1/R \leq 0.5$. Figure 8.46a shows that the curvature increases with deviatoric strain. The equation of the circular arcs is

$$\left[\cos 3\alpha_\sigma + \sqrt{(0.5R^2 - 1)}\right]^2 + \left[\cos 3\alpha_\varepsilon - \sqrt{(0.5R^2 - 1)}\right]^2 = R^2 \tag{8.52}$$

which can be rearranged to give

$$\cos 3\alpha_\sigma = \left\{R^2 - \left[\cos 3\alpha_\varepsilon - \sqrt{(0.5R^2 - 1)}\right]^2\right\}^{1/2} - \sqrt{(0.5R^2 - 1)} \tag{8.53}$$

Since, for radial deviatoric strain paths, the directions of strain and strain-rate are the same, the above equation relates the direction of the deviatoric stress vector to the direction of the deviatoric strain-rate vector and thereby provides a deviatoric flow rule.

The size of the stress response envelopes can be defined by $\left(1 - I_3^*\right)$ which is plotted against $1/R$ in Fig. 8.46b. The figure clearly indicates a power law relationship that has the form

$$\left(1 - I_3^*\right) = (1/R)^{3/2} \tag{8.54}$$

which provides a scaling law. There is increasing evidence from both simulations and experiments that, for granular media, failure envelopes and stress response envelopes in general are accurately defined by Lade surfaces. The size of such surfaces, however, is significantly greater for real granular materials due to particle shape effects. Consequently, for real granular material, it is expected that the scaling law needs to be modified to

$$\left(1 - I_3^*\right) = A(1/R)^{3/2} \text{ with } A \geq 1 \tag{8.55}$$

The work reported in this subsection is an excellent demonstration of the power of DEM and its ability to contribute to our understanding of the behaviour of granular material under complex states of stress and strain. It is the opinion of the author that the simulated results presented above are qualitatively generic and that particle shape will only affect the magnitude of the various parameters but will not affect the form of the various relationships. Attempts to validate the findings by performing experiments in general 3D stress space are extremely difficult due to the inability to prepare exact replicas of the initial particle system. In contrast, DEM has the significant advantage that an infinite number of different tests can be simulated for exactly the same initial sample conditions. However, the work reported in this subsection indicates that, in terms of general 3D stress space, there is much more work that needs to be done.

References

Bathurst, R.J., Rothenburg, L.: Micromechanical aspects of isotropic granular assemblies with linear contact interactions. J. Appl. Mech. **55**, 17–23 (1988)

Bishop, A.W.: The strength of soils as engineering materials. Geotechnique **16**, 91–128 (1966)

Bolton, M.D.: The strength and dilatancy of sands. Geotechnique **36**, 65–78 (1986)

Calvetti, F., Viggiani, G., Tamagnini, C.: A numerical investigation of the incremental behaviour of granular soils. Riv. Ital. Geotecnica **37**, 11–29 (2003)

Cundall, P.A., Strack, O.D.L.: A discrete numerical model for granular assemblies. Geotechnique **29**, 47–65 (1979)

Dafalias, Y.F., Popov, E.P.: Plastic internal variables formalism of cyclic plasticity. J. Appl. Mech. **43**, 645–651 (1976)

Dantu, P.: A contribution to the mechanical and geometrical study of non-cohesive masses. In: Proc. 4th Int. Conf. on Soil Mechanics and Foundation Engineering, pp. 144–148. Butterworths, London (1957)

de Josselin de Jong, G., Verruijt, A.: Étude photo-élastique d'un empilement de disques. Cah. Groupe Fr. Rheologie **2**, 73–86 (1969)

Deresiewicz, H.: Stress-strain relations for a simple model of a granular medium. J. Appl. Mech. **25**, 402–406 (1958)

Digby, P.J.: The effective elastic moduli of porous granular rocks. J. Appl. Mech. **48**, 803–808 (1981)

Drescher, A., de Josselin de Jong, G.: Photoelastic verification of a mechanical model for the flow of a granular material. J. Mech. Phys. Solids **20**, 337–351 (1972)

Duffy, J.: A differential stress-strain relation for the hexagonal close-packed array of elastic spheres. J. Appl. Mech. **26**, 88–94 (1959)

Duffy, J., Mindlin, R.D.: Stress strain relations and vibrations of a granular medium. J. Appl. Mech. **24**, 585–593 (1957)

Dyer, M.R., Milligan, G.W.E.: A photoelastic investigation of the interaction of a cohesionless soil with reinforcement placed at different orientations. In: Proc. Int. Conf. In Situ Soil Rock Reinf., pp. 257–262. Presse Ponts et Chaussees, Paris (1984) (Fig. 7, p. 260)
Hicher, P.Y., Chang, C.S.: Evaluation of two homogenization techniques for modelling the elastic behaviour of granular materials. J. Eng. Mech. ASCE **131**, 1184–1194 (2005)
Hill, R.: The Mathematical Theory of Plasticity. Oxford University Press, Oxford (1950)
Horne, M.R.: The behaviour, rigid, cohesionless particles, III. Proc. R. Soc. A **310**, 21–34 (1969)
Joer, H.A., Lanier, J., Desruees, J., Flavigny, E.: $1\gamma 2\varepsilon$: a new shear apparatus to study the behaviour of granular materials. Geotech. Test. J. ASTM **15**, 129–137 (1992)
Kruyt, N.P., Rotheburg, L.: Micromechanical bounds for the effective elastic moduli of granular materials. Int. J. Solids Struct. **39**, 311–324 (2002)
Kuhn, M.R.: Structure deformation in granular materials. Mech. Mater. **31**, 407–449 (1999)
Lade, P.V., Duncan, J.M.: Elastoplastic stress-strain theory for cohesionless soil. J. Geotech. Eng. ASCE **101**, 1037–1053 (1975)
Leussink, H., Wittke, W.: Difference in triaxial and plane strain shear strength. In: Symp. Lab. Shear Testing Soils. STP 361, pp. 77–89. American Society for Testing and Materials, Ottawa (1963)
Mandl, G., Fernandez-Luque, R.: Fully developed plastic shear flow of granular materials. Geotechnique **20**, 277–307 (1970)
Matsuoka, H., Nakai, T.: Stress-deformation and strength characteristics of soil under three different principal stresses. Proc. Jpn Soc. Civ. Eng. **232**, 59–70 (1974)
Ng, T.T.: Shear strength of assemblies of ellipsoidal particles. Geotechnique **54**, 659–669 (2004)
Oda, M., Konishi, J.: Microscopic deformation mechanism of granular material in simple shear. Soils Found. **14**, 25–38 (1974)
Prager, W.: The theory of plasticity: a survey of recent achievements. Proc. Inst. Mech. Eng. **169**, 41–57 (1955)
Radjai, F., Wolf, D.E., Roux, S., Jean, M., Moreau, J.J.: Force networks in dense granular media. In: Behringer, R.P., Jenkins, J.T. (eds.) Powders & Grains 97, pp. 211–214. Balkema, Rotterdam (1997)
Rennie, B.C.: On the strength of sand. J. Aust. Math. Soc. **1**, 71–79 (1959)
Rothenburg, L., Bathurst, R.J.: Analytical study of induced anisotropy in idealised granular materials. Geotechnique **39**, 601–614 (1989)
Rowe, P.W.: The stress-dilatancy relation for static equilibrium of an assembly of particles in contact. Proc. R. Soc. A **269**, 500–527 (1962)
Schofield, A.N., Wroth, C.P.: Critical State Soil Mechanics. McGraw-Hill, London (1968)
Skinner, A.: A note on the influence of interparticle friction on the shearing strength of a random assembly of spherical particles. Geotechnique **19**, 150–157 (1969)
Taylor, D.W.: Fundamentals of Soil Mechanics. Wiley, New York (1948)
Thornton, C.: Force transmission in granular media. KONA Powder Part. **15**, 81–90 (1997)
Thornton, C.: Numerical simulations of deviatoric shear deformation of granular media. Geotechnique **50**, 43–53 (2000)
Thornton, C., Antony, S.J.: Quasi-static deformation of particulate media. Phil. Trans. R. Soc. Lond. A **356**, 2763–2782 (1998)
Thornton, C., Barnes, D.J.: On the mechanics of granular material. In: Vermeer, P.A., Luger, H.J. (eds.) IUTAM Symp. on Deformation and Failure of Granular Materials, pp. 69–77. Balkema, Rotterdam (1982)
Thornton, C., Barnes, D.J.: Computer simulated deformation of compact granular assemblies. Acta Mech. **64**, 45–61 (1986)
Thornton, C., Sun, G.: Axisymmetric compression of 3D polydisperse systems of spheres. In: Thornton, C. (ed.) Powders & Grains 93, pp. 129–134. Balkema, Rotterdam (1993)
Thornton, C., Zhang, L.: Numerical simulations of the direct shear test. Chem. Eng. Technol. **26**, 153–156 (2003)

Thornton, C., Zhang, L.: On the elastic response of granular media. In: Garcia-Rojo, R., Herrmann, H.J., McNamara, S. (eds.) Powders and Grains 2005, pp. 235–238. Balkema, Leiden (2005)

Thornton, C., Zhang, L.: A numerical examination of shear banding and simple shear non-coaxial flow rules. Philos. Mag. **86**, 3425–3452 (2006)

Thornton, C., Zhang, L.: On the evolution of stress and microstructure during general 3D deviatoric straining of granular media. Geotechnique **60**, 333–341 (2010)

Vardoulakis, I., Sulem, J.: Bifurcation Analysis in Geomechanics. Blackie Academic & Professional, London (1995) (Fig. 6.1.1)

Wakabayshi, T.: Photoelastic method for determination of stress in powdered mass. In: Proc. 7th Japan National Congress on Applied Mechanics, pp. 153–158 (1957)

Walton, K.: The effective elastic moduli of a random packing of spheres. J. Mech. Phys. Solids **35**, 213–226 (1987)

Williams, J.R., Rege, N.: The development of circulation cell structures in granular materials undergoing compression. Powder Technol. **90**, 187–194 (1997)

Zhang, L., Thornton, C.: A numerical examination of the direct shear test. Geotechnique **57**, 343–354 (2007)

Ziegler, H.: A modification of Prager's hardening rule. Q. J. Appl. Math **17**, 55–65 (1959)

FSC
www.fsc.org
MIX
Papier | Fördert
gute Waldnutzung
FSC® C083411

Kolibri 360 GmbH
Ferdinand-Jühlke-Straße 7
99095 Erfurt, Deutschland
produktsicherheit@kolibri360.de